Traffic System
Design Handbook

Telecommunications Handbook Series
Whitham D. Reeve, *Series Editor*

The *Telecommunications Handbook Series* is designed to provide the engineer and technical practitioner with working information in the three basic fields of telecommunications: inside plant, outside plant, and administration and regulatory. This integrated series of handbooks provides practical information on the link between field experience and formal telecommunication industry standards and practices. These books are essential tools for engineers and technical practitioners who require day-to-day engineering and technical information on telecommunication systems.

Other proposed and forthcoming books include coverage on:

- Introduction to telecommunications
- Grounding methods and measurements
- Telecommunication protection
- Power system design
- Fiber-in-the-loop design
- Telecommunication construction methods
- Satellite earth station design
- Terrestrial radio system design
- Telecommunications regulatory
- Personal communications
- Switching and networking

If you are interested in becoming an author, contributor, or reviewer of a book in this series, or if you would like additional information about forthcoming titles, please contact:

Whitham D. Reeve
Series Editor, Telecommunications Handbook Series
PO Box 190225
Anchorage, Alaska 99519-0225

Traffic System
Design Handbook
Timesaving Telecommunication
Traffic Tables and Programs

James R. Boucher, P.E.

A volume in the Telecommunications Handbook Series,
Whitham D. Reeve, *Series Editor*

 IEEE
PRESS

 Reviewed and endorsed by the
IEEE Communications Society

The Institute of Electrical and Electronics Engineers, Inc., New York

©1993 by the Institute of Electrical and Electronics Engineers, Inc.
345 East 47th Street, New York, NY 10017-2394

Printed in the United States of America

10   9   8   7   6   5   4   3   2   1

ISBN 0-7803-0428-4

IEEE Order Number: PP0325-1

Library of Congress Cataloging-in-Publication Data
Boucher, James R. (date)
    Traffic system design handbook : timesaving telecommunication
traffic tables and·programs / James R. Boucher.
    p.   cm.—(Telecommunication handbook)
    Includes bibliographical reference and index.
    ISBN 0-7803-0428-4
    1. Telecommunication—Traffic   2. Telecommunication—Switching
systems.   I. Title.   II. Series.
TK5102.5.B683 1993
621.382—dc20
                                                                    92-30770
                                                                    CIP

To my wife, Barbara

Contents

Contents

Preface

This handbook is intended for engineers, technicians, and others who specify, design, test, operate, or maintain telephone systems and networks. It is the handbook I would have referred to most often if it had been available earlier in my career of engineering telecommunication switching and control systems for domestic, international, and military applications. Instead, I amassed a considerable traffic-theory library, including many formulas, tables, curves, and charts. From these I developed a family of timesaving computer programs that greatly simplified my traffic system design tasks.

The most commonly used of these diverse design resources, with examples of their use in practical applications, are now available in a single volume. Where typical values of parameters are given, they are indicative of industry practice—individual companies may use different values based on engineering requirements or policy. For those new to traffic system design, introductory information, definitions of common terms and abbreviations, and an extensive bibliography are included.

I would be remiss if I did not acknowledge and thank the staff who developed and conducted the GTE Traffic Engineering School held at Norwalk, CT in 1974. The notes for that course introduced me to traffic system design and became the basis for the course I teach at Northeastern University. I must also thank my colleagues, Armand L. Damien and John C. Rothwell, for their critical review of the manuscript for this handbook.

To Louis K. Pollen, my mentor, colleague, and friend, I extend my sincere appreciation for his guidance through the years, and my best wishes for his retirement.

In addition, I am indebted to Dudley R. Kay, IEEE Press Executive Editor, for encouraging me to take on this project; Whitham D. Reeve, Telecommunications Handbook Series Editor, for his sound advice and comments along the way; and to Denise Gannon, Production Supervisor, and all the staff at IEEE Press who helped bring this handbook to fruition.

J. R. Boucher

List of Tables

1

Traffic System Design Overview

Traffic system design is a process that considers the entire telecommunication system and the interrelationship of its components. Total system and subsystem performance (i.e., service) objectives are specified, and conflicts are resolved to achieve an optimum configuration. Therefore, traffic system design ensures the cost-effective dimensioning of switching and transmission equipment (traffic-handling resources or servers) to provide the required service objectives (grade of service) economically. Telephone traffic (teletraffic) theory—drawing on many disciplines including electronics, mathematics, statistics, probability, queuing theory, reliability, and economics—is at the heart of traffic system design.

1.1 TRAFFIC UNITS

Traffic units are a measure of traffic intensity, the average traffic density during a one-hour period. The international unit of traffic intensity is the Erlang,* where one Erlang represents a circuit occupied for one hour.

---

* Named for A.K. Erlang, the father of telephone traffic theory [Brockmeyer, 1948].

The Erlang defines the efficiency (percent occupancy) of a traffic resource and represents the total time in hours to carry all calls. It is the traffic unit used exclusively in classic traffic theory.

In the North American public switched telephone network (PSTN), the standard traffic unit is the unit call (UC), which is expressed in seconds. The UC is defined in centum-call-seconds (CCS) or more commonly, hundred-call-seconds. Equation 1.1 gives the relationship between Erlangs and CCS. Table 1-1 is an Erlang-to-CCS conversion chart for selected traffic levels up to 200 Erlangs (7200 CCS).

$$1 \text{ Erlang} = 1 \text{ call-hour} = 3600 \text{ call-seconds} = 36 \text{ CCS} \qquad (1.1)$$

1.2 TRAFFIC CALCULATIONS

Before common-equipment pools such as trunk groups, signaling registers, and operator positions can be dimensioned, their busy-hour traffic intensities must be determined. Trunks are assigned to serve calls on an immediate basis and are held for the duration of the call. Signaling registers, operator positions, and similar servers normally serve calls on a delayed basis and are held only long enough to serve their specific functions.

1.2.1 Trunk-Group Traffic

Routing plans specify a mix of direct-route and alternate-route trunk groups to provide least-cost routing of interswitch traffic through the network. The selected routing technique determines, to some extent, the level of traffic offered to each trunk group. Offered trunk-group traffic is the total of all traffic offered to the group. If the trunk group were large enough, it would carry all offered traffic but such a trunk group probably would not be economical. Instead, trunk groups are engineered to block a fraction of the offered busy-hour traffic, typically one to ten percent.

Table 1-1. Traffic-Unit Conversion Chart

Erlangs	CCS	Erlangs	CCS	Erlangs	CCS	Erlangs	CCS	Erlangs	CCS
0.05	1.8	2.05	73.8	4.05	145.8	6.05	217.8	8.05	289.8
0.10	3.6	2.10	75.6	4.10	147.6	6.10	219.6	8.10	291.6
0.15	5.4	2.15	77.4	4.15	149.4	6.15	221.4	8.15	293.4
0.20	7.2	2.20	79.2	4.20	151.2	6.20	223.2	8.20	295.2
0.25	9.0	2.25	81.0	4.25	153.0	6.25	225.0	8.25	297.0
0.30	10.8	2.30	82.8	4.30	154.8	6.30	226.8	8.30	298.8
0.35	12.6	2.35	84.6	4.35	156.6	6.35	228.6	8.35	300.6
0.40	14.4	2.40	86.4	4.40	158.4	6.40	230.4	8.40	302.4
0.45	16.2	2.45	88.2	4.45	160.2	6.45	232.2	8.45	304.2
0.50	18.0	2.50	90.0	4.50	162.0	6.50	234.0	8.50	306.0
0.55	19.8	2.55	91.8	4.55	163.8	6.55	235.8	8.55	307.8
0.60	21.6	2.60	93.6	4.60	165.6	6.60	237.6	8.60	309.6
0.65	23.4	2.65	95.4	4.65	167.4	6.65	239.4	8.65	311.4
0.70	25.2	2.70	97.2	4.70	169.2	6.70	241.2	8.70	313.2
0.75	27.0	2.75	99.0	4.75	171.0	6.75	243.0	8.75	315.0
0.80	29.8	2.80	100.8	4.80	172.8	6.80	244.8	8.80	316.8
0.85	30.6	2.85	102.6	4.85	174.6	6.85	246.6	8.85	318.6
0.90	32.4	2.90	104.4	4.90	176.4	6.90	248.4	8.90	320.4
0.95	34.2	2.95	106.2	4.95	178.2	6.95	250.2	8.95	322.2
1.00	36.0	3.00	108.0	5.00	180.0	7.00	252.0	9.00	324.0
1.05	37.8	3.05	109.8	5.05	181.8	7.05	253.9	9.05	325.8
1.10	39.6	3.10	111.6	5.10	183.6	7.10	255.6	9.10	327.6
1.15	41.4	3.15	113.4	5.15	185.4	7.15	257.4	9.15	329.4
1.20	43.2	3.20	115.2	5.20	187.2	7.20	259.2	9.20	331.2
1.25	45.0	3.25	117.0	5.25	189.0	7.25	261.0	9.25	333.0
1.30	46.8	3.30	118.8	5.30	190.8	7.30	262.8	9.30	334.8
1.35	48.6	3.35	120.6	5.35	192.6	7.35	264.6	9.35	336.6
1.40	50.4	3.40	122.4	5.40	194.4	7.40	266.4	9.40	338.4
1.45	52.2	3.45	124.2	5.45	196.2	7.45	268.2	9.45	340.2
1.50	54.0	3.50	126.0	5.50	198.0	7.50	270.0	9.50	342.0
1.55	55.8	3.55	127.8	5.55	199.8	7.55	271.8	9.55	343.8
1.60	57.6	3.60	129.6	5.60	201.6	7.60	273.6	9.60	345.6
1.65	59.4	3.65	131.4	5.65	203.4	7.65	275.4	9.65	347.4
1.70	61.2	3.70	133.2	5.70	205.2	7.70	277.2	9.70	349.2
1.75	63.0	3.75	135.0	5.75	207.0	7.75	279.0	9.75	351.0
1.80	64.8	3.80	136.8	5.80	208.8	7.80	280.8	9.80	352.8
1.85	66.6	3.85	138.6	5.85	210.6	7.85	282.6	9.85	354.6
1.90	68.4	3.90	140.4	5.90	212.4	7.90	284.4	9.90	356.4
1.95	70.2	3.95	142.2	5.95	214.2	7.95	285.2	9.95	358.2
2.00	72.0	4.00	144.0	6.00	216.0	8.00	288.0	10.00	360.0

(*table continues*)

Table 1-1. Traffic-Unit Conversion Chart (*Continued*)

Erlangs	CCS	Erlangs	CCS	Erlangs	CCS	Erlangs	CCS	Erlangs	CCS
10.1	363.6	14.1	507.6	18.1	651.6	22.1	795.6	26.1	939.6
10.2	367.2	14.2	511.2	18.2	654.2	22.2	799.2	26.2	943.2
10.3	370.8	14.3	514.8	18.3	658.8	22.3	802.8	26.3	946.8
10.4	374.4	14.4	518.4	18.4	662.4	22.4	806.4	26.4	950.4
10.5	378.0	14.5	522.0	18.5	666.0	22.5	810.0	26.5	954.0
10.6	381.6	14.6	525.6	18.6	669.6	22.6	813.6	26.6	957.6
10.7	385.2	14.7	529.2	18.7	673.2	22.7	817.2	26.7	961.2
10.8	388.8	14.8	532.8	18.8	676.8	22.8	820.8	26.8	964.8
10.9	392.4	14.9	536.4	18.9	680.4	22.9	824.4	26.9	968.4
11.0	396.0	15.0	540.0	19.0	684.0	23.0	828.0	27.0	972.0
11.1	399.6	15.1	543.6	19.1	687.6	23.1	831.6	27.1	975.6
11.2	403.2	15.2	547.2	19.2	691.2	23.2	835.2	27.2	979.2
11.3	406.8	15.3	550.8	19.3	694.8	23.3	838.8	27.3	982.8
11.4	410.4	15.4	554.4	19.4	698.4	23.4	842.4	27.4	986.4
11.5	414.0	15.5	558.0	19.5	702.0	23.5	846.0	27.5	990.0
11.6	417.6	15.6	561.6	19.6	705.6	23.6	849.6	27.6	993.6
11.7	421.2	15.7	565.2	19.7	709.2	23.7	853.2	27.7	997.2
11.8	424.8	15.8	568.8	19.8	712.8	23.8	856.8	27.8	1000.8
11.9	428.4	15.9	572.4	19.9	716.2	23.9	860.2	27.9	1004.2
12.0	432.0	16.0	576.0	20.0	720.0	24.0	864.0	28.0	1008.0
12.1	431.6	16.1	579.6	20.1	723.6	24.1	867.6	28.1	1011.6
12.2	439.2	16.2	583.2	20.2	727.2	24.2	871.2	28.2	1015.2
12.3	442.8	16.3	586.8	20.3	730.8	24.3	874.8	28.3	1018.8
12.4	446.4	16.4	590.4	20.4	734.4	24.4	878.4	28.4	1022.4
12.5	450.0	16.5	594.0	20.5	738.0	24.5	882.0	28.5	1026.0
12.6	453.6	16.6	597.6	20.6	741.6	24.6	885.6	28.6	1029.6
12.7	457.2	16.7	601.2	20.7	745.2	24.7	889.2	28.7	1033.2
12.8	460.8	16.8	604.8	20.8	748.8	24.8	892.8	28.8	1036.8
12.9	464.4	16.9	608.4	20.9	752.4	24.9	896.4	28.9	1040.4
13.0	468.0	17.0	612.0	21.0	756.0	25.0	900.0	29.0	1044.0
13.1	471.6	17.1	615.6	21.1	759.6	25.1	903.6	29.1	1047.6
13.2	475.2	17.2	619.2	21.2	763.2	25.2	907.2	29.2	1051.2
13.3	478.8	17.3	622.8	21.3	766.8	25.3	910.8	29.3	1054.8
13.4	482.4	17.4	626.4	21.4	770.4	25.4	914.4	29.4	1058.4
13.5	486.0	17.5	630.0	21.5	774.0	25.5	918.0	29.5	1062.0
13.6	489.6	17.6	633.6	21.6	777.6	25.6	921.6	29.6	1065.6
13.7	493.2	17.7	637.2	21.7	781.2	25.7	925.2	29.7	1069.2
13.8	496.8	17.8	640.8	21.8	784.8	25.8	928.8	29.8	1072.8
13.9	500.4	17.9	644.4	21.9	788.4	25.9	932.4	29.9	1076.2
14.0	504.0	18.0	648.0	22.0	792.0	26.0	936.0	30.0	1080.0

(*table continues*)

Table 1-1. Traffic-Unit Conversion Chart (*Continued*)

Erlangs	CCS	Erlangs	CCS	Erlangs	CCS	Erlangs	CCS	Erlangs	CCS
30.1	1083.6	34.1	1227.6	38.1	1371.6	42.1	1515.6	46.1	1659.6
30.2	1087.2	34.2	1231.2	38.2	1375.2	42.2	1519.2	46.2	1663.2
30.3	1090.8	34.3	1234.8	38.3	1378.8	42.3	1522.8	46.3	1666.8
30.4	1094.2	34.4	1238.4	38.4	1382.4	42.4	1526.4	46.7	1670.4
30.5	1098.0	34.5	1242.0	38.5	1386.0	42.5	1530.0	46.5	1674.0
30.6	1101.6	34.6	1245.6	38.6	1389.6	42.6	1533.6	46.6	1677.6
30.7	1105.2	34.7	1249.2	38.7	1393.2	42.7	1537.2	46.7	1681.2
30.8	1108.8	34.8	1252.8	38.8	1396.8	42.8	1540.8	46.8	1684.8
30.9	1112.4	34.9	1256.4	38.9	1400.4	42.9	1544.4	46.9	1688.4
31.0	1116.0	35.0	1260.0	39.0	1404.0	43.0	1548.0	47.0	1692.0
31.1	1119.6	35.1	1263.6	39.1	1407.6	43.1	1551.6	47.1	1695.6
31.2	1123.2	35.2	1267.2	39.2	1411.2	43.2	1555.2	47.2	1699.2
31.3	1126.8	35.3	1270.8	39.3	1414.8	43.3	1558.8	47.3	1702.8
31.4	1130.4	35.4	1274.4	39.4	1418.4	43.4	1562.4	47.4	1706.4
31.5	1134.0	35.5	1278.0	39.5	1422.0	43.5	1566.0	47.5	1710.0
31.6	1137.6	35.6	1281.6	39.6	1425.6	43.6	1569.6	47.6	1713.6
31.7	1141.2	35.7	1285.2	39.7	1429.2	43.7	1573.2	47.7	1717.4
31.8	1144.8	35.8	1288.8	39.8	1432.8	43.8	1576.8	47.8	1720.8
31.9	1148.4	35.9	1292.4	39.9	1436.4	43.9	1580.4	47.9	1724.4
32.0	1152.0	36.0	1296.0	40.0	1440.0	44.0	1584.0	48.0	1728.0
32.1	1155.6	36.1	1299.6	40.1	1443.6	44.1	1587.6	48.1	1731.6
32.2	1159.2	36.2	1303.2	40.2	1447.2	44.2	1591.2	48.2	1735.2
32.3	1162.8	36.3	1306.8	40.3	1450.8	44.3	1594.8	48.3	1738.8
32.4	1166.4	36.4	1310.4	40.4	1454.4	44.4	1598.4	48.4	1742.4
32.5	1170.0	36.5	1314.0	40.5	1458.0	44.5	1602.0	48.5	1746.0
32.6	1173.6	36.6	1317.6	40.6	1461.6	44.6	1605.6	48.6	1749.6
32.7	1177.2	36.7	1321.2	40.7	1465.2	44.7	1609.2	48.7	1753.2
32.8	1180.8	36.8	1324.8	40.8	1468.8	44.8	1612.8	48.8	1756.8
32.9	1184.4	36.9	1328.4	40.9	1472.4	44.9	1616.4	48.9	1760.4
33.0	1188.0	37.0	1332.0	41.0	1476.0	45.0	1620.0	49.0	1764.0
33.1	1191.6	37.1	1335.6	41.1	1479.6	45.1	1623.6	49.1	1767.6
33.2	1195.2	37.2	1339.2	41.2	1483.2	45.2	1627.2	49.2	1771.2
33.3	1198.8	37.3	1342.8	41.3	1486.8	45.3	1630.8	49.3	1774.8
33.4	1202.4	37.4	1346.4	41.4	1490.4	45.4	1634.4	49.4	1778.4
33.5	1206.0	37.5	1350.0	41.5	1494.0	45.5	1638.0	49.5	1782.0
33.6	1209.6	37.6	1353.6	41.6	1497.6	45.6	1641.6	49.6	1785.6
33.7	1213.2	37.7	1357.2	41.7	1501.2	45.7	1645.2	49.7	1789.2
33.8	1216.8	37.8	1360.8	41.8	1504.8	45.8	1648.8	49.8	1792.8
33.9	1220.4	37.9	1364.4	41.9	1508.4	45.9	1652.4	49.9	1796.4
34.0	1224.0	38.0	1368.0	42.0	1512.0	46.0	1656.0	50.0	1800.0

(*table continues*)

Table 1-1. Traffic-Unit Conversion Chart (*Continued*)

Erlangs	CCS	Erlangs	CCS	Erlangs	CCS	Erlangs	CCS	Erlangs	CCS
50.5	1818	70.5	2538	90.5	3258	121.0	4356	161.0	5796
51.0	1836	71.0	2556	91.0	3276	122.0	4392	162.0	5832
51.5	1854	71.5	2574	91.5	3294	123.0	4428	163.0	5868
52.0	1872	72.0	2592	92.0	3312	124.0	4464	164.0	5904
52.5	1890	72.5	2610	92.5	3330	125.0	4500	165.0	5940
53.0	1908	73.0	2628	93.0	3348	126.0	4536	166.0	5976
53.5	1926	73.5	2646	93.5	3366	127.0	4572	167.0	6012
54.0	1944	74.0	2664	94.0	3384	128.0	4608	168.0	6048
54.5	1962	74.5	2682	94.5	3402	129.0	4644	169.0	6084
55.0	1980	75.0	2700	95.0	3420	130.0	4680	170.0	6120
55.5	1998	75.5	2718	95.5	3438	131.0	4716	171.0	6156
56.0	2016	76.0	2736	96.0	3456	132.0	4752	172.0	6192
56.5	2034	76.5	2754	96.5	3474	133.0	4788	173.0	6228
57.0	2052	77.0	2772	97.0	3492	134.0	4824	174.0	6264
57.5	2070	77.5	2790	97.5	3510	135.0	4860	175.0	6300
58.0	2088	78.0	2808	98.0	3528	136.0	4896	176.0	6336
58.5	2106	78.5	2826	98.5	3546	137.0	4932	177.0	6372
59.0	2124	79.0	2844	99.0	3564	138.0	4968	178.0	6408
59.5	2142	79.5	2862	99.5	3582	139.0	5004	179.0	6444
60.0	2160	80.0	2880	100.0	3600	140.0	5040	180.0	6480
60.5	2178	80.5	2898	101.0	3636	141.0	5076	181.0	6516
61.0	2196	81.0	2916	102.0	3672	142.0	5112	182.0	6552
61.5	2214	81.5	2934	103.0	3708	143.0	5148	183.0	6588
62.0	2232	82.0	2952	104.0	3744	144.0	5184	184.0	6624
62.5	2250	82.5	2970	105.0	3780	145.0	5220	185.0	6660
63.0	2268	83.0	2988	106.0	3816	146.0	5256	186.0	6696
63.5	2286	83.5	3006	107.0	3852	147.0	5292	187.0	6732
64.0	2304	84.0	3024	108.0	3888	148.0	5328	188.0	6768
64.5	2322	84.5	3042	109.0	3924	149.0	5364	189.0	6804
65.0	2340	85.0	3060	110.0	3960	150.0	5400	190.0	6840
65.5	2358	85.5	3078	111.0	3996	151.0	5436	191.0	6876
66.0	2376	86.0	3096	112.0	4032	152.0	5472	192.0	6912
66.5	2394	86.5	3114	113.0	4068	153.0	5508	193.0	6948
67.0	2412	87.0	3132	114.0	4104	154.0	5544	194.0	6984
67.5	2430	87.5	3150	115.0	4140	155.0	5580	195.0	7020
68.0	2448	88.0	3168	116.0	4176	156.0	5616	196.0	7056
68.5	2466	88.5	3186	117.0	4212	157.0	5652	197.0	7092
69.0	2484	89.0	3204	118.0	4248	158.0	5688	198.0	7128
69.5	2502	89.5	3222	119.0	4284	159.0	5724	199.0	7164
70.0	2520	90.0	3240	120.0	4320	160.0	5760	200.0	7200

Figure 1-1 can be used to facilitate an understanding of traffic routing terms. Interswitch traffic is routed over the primary route trunk group provided there are idle trunks available in the group. In an alternate-routing system, blocked trunk-group traffic overflows to other alternate-route trunk groups or to final-route trunk groups as indicated by the curved arrows. Trunk groups provided with alternate routes are often referred to as *high-usage trunk groups*. Final-route trunk groups do not have alternate routes; therefore, blocked traffic in a final-route trunk group is lost.

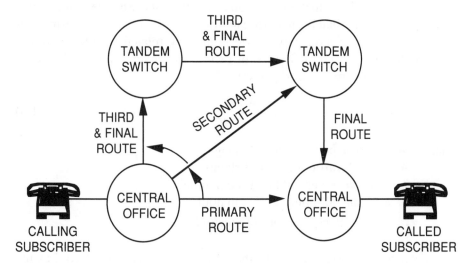

Figure 1-1.   Interswitch Trunk Traffic Routing Diagram

Trunk-group traffic is the product of the number and duration of calls handled by the group. Equation 1.2 can be used to calculate trunk-group traffic, expressed in Erlangs.

$$A = N \cdot T_c \tag{1.2}$$

where A = Offered traffic in Erlangs
N = Number of calls during the busy hour
T_c = Mean call-holding time in hours

Number of calls refers to the total number of calls offered to the trunk group. Call-holding time is the total elapsed time between seizure of a trunk to serve the call and its subsequent release. The mean call-holding time is the arithmetic average of all call-holding times, expressed in hours.

Example 1-1

Determine the traffic in Erlangs and CCS for a trunk group carrying 1000 calls during the busy hour with an average call-holding time of 3 minutes.

A = (1000 calls/hour)(3 min/call)(1 hour/60 min) = 50 Erlangs
(50 Erl)(36 CCS/Erl) = 1800 CCS

1.2.2 Server-Pool Traffic

Server pools are groups of traffic resources, such as signaling registers and operator positions, that are used on a shared basis. Service requests that cannot be met immediately are placed in a queue and served on a first-in, first-out (FIFO) basis. Server-pool traffic is directly related to offered traffic, server-holding time, and call-attempt factor, and inversely related to call-holding time as expressed in Equation 1.3.

$$A_S = \frac{A_T \cdot T_s \cdot C}{T_C} \tag{1.3}$$

where A_S = Server-pool traffic in Erlangs
A_T = Total traffic served in Erlangs
T_S = Mean server-holding time in hours
T_C = Mean call-holding time in hours
C = Call-attempt factor (dimensionless)

Total traffic served refers to the total offered traffic that requires the services of the specific server pool for some portion of the call. For example, a dual-tone multifrequency (DTMF) receiver pool is dimensioned to serve only the DTMF tone-dialing portion of total switch traffic generated by DTMF signaling sources. Table 1-2 presents representative server-holding times for typical signaling registers as a function of the number of digits received or sent.

Table 1-2. Typical Signaling Register Holding Times in Seconds

Signaling Register	Number of Digits Received or Sent				
	1	4	7	10	11
Local Dial-Pulse (DP) Receiver	3.7	8.3	12.8	17.6	19.1
Local DTMF Receiver	2.3	5.2	8.1	11.0	12.0
Incoming MF Receiver	1.0	1.4	1.8	2.2	2.3
Outgoing MF Sender	1.5	1.9	2.3	2.8	3.0

The mean server-holding time is the arithmetic average of all server-holding times for the specific server pool. Equation 1.4 is a general equation to calculate mean server-holding time for calls with different holding-time characteristics.

$$T_S = a{\cdot}T_1 + b{\cdot}T_2 + \cdots + k{\cdot}T_n \qquad (1.4)$$

where
$$T_S = \text{Mean server-holding time in hours}$$
$$T_1, T_2, \cdots, T_n = \text{Individual server-holding times in hours}$$
$$a, b, \cdots, k = \text{Fractions of total traffic served } (a + b + \cdots + k = 1)$$

Example 1-2

Determine the mean DTMF receiver-holding time for a central office (CO) where subscribers dial local calls using a 7-digit number and toll calls using an 11-digit number. Assume that 70 percent of the calls are local calls, the remainder are toll calls, and that the typical signaling register holding times of Table 1-2 are applicable.

$$T_S = (0.7)(8.1 \text{ sec}) + (0.3)(12.0 \text{ sec}) = 9.27 \text{ sec}$$

Call-attempt factors are dimensionless numbers that adjust offered traffic intensity to compensate for call attempts that do not result in completed calls. Therefore, call-attempt factors are inversely proportional to the fraction of completed calls as defined in Equation 1.5.

$$C = \frac{1}{k} \qquad (1.5)$$

where C = Call-attempt factor (dimensionless)
  k = Fraction of calls completed (decimal fraction)

Example 1-3

Table 1-3 presents representative subscriber call-attempt dispositions based on empirical data amassed in the North American PSTN. Determine the call-attempt factor for these data, where 70.7 percent of the calls were completed ($k = 0.707$).

$$C = \frac{1}{k} = \frac{1}{0.707} = 1.414$$

Table 1-3. Typical Call-Attempt Dispositions

Call-Attempt Disposition	Percentage
Call was completed	70.7
Called subscriber did not answer	12.7
Called subscriber line was busy	10.1
Call abandoned without system response	2.6
Equipment blockage or failure	1.9
Customer dialing error	1.6
Called directory number changed or disconnected	0.4

Example 1-4

Using Equation 1.3, determine the server-pool traffic in CCS and Erlangs for the DTMF receivers of Example 1-2, assuming total offered busy-hour subscriber traffic of 2000 CCS, a call-attempt factor of 1.5, and a mean call-holding time of 3 minutes (180 seconds).

$$A_S = (2000 \text{ CCS}) \, (1.5) \frac{(9.27 \text{ sec})}{(180 \text{ sec})} = 154.5 \text{ CCS}$$

$$(154.5 \text{ CCS}) \, \frac{(1 \text{ Erl})}{(36 \text{ CCS})} = 4.29 \text{ Erlangs}$$

1.3 TRAFFIC ASSUMPTIONS

Traffic formulas are based on a set of assumptions regarding the behavior of traffic and its sources. These assumptions are not always precisely true. If variations from these assumptions are small or known to have little effect, however, they can be used with confidence.

1.3.1 General Assumptions

The following assumptions are applicable to traffic formulas in general:

- The system is in statistical equilibrium.
- Connection and disconnection of sources to servers occur instantaneously.
- The anticipated traffic density is the same for all sources.
- Busy sources initiate no calls.
- Every source has equal access to every server (full availability).

- The number of busy servers in a group is equal to the number of busy sources in its group of sources.

1.3.2 Number of Sources

The number of sources that can originate calls affects the service these sources can expect to obtain. As the number of sources increases, the effect of adding more sources diminishes. Eventually, a point is reached where there is negligible difference in the probability of congestion regardless of how many new sources are added. It is this point that distinguishes between finite and infinite sources. Traffic formulas for applications where the number of sources in relation to the number of servers is very large assume infinite sources (worst case for blocking). This simplifies the mathematics and minimizes the number of required tables.

1.3.3 Disposition of Blocked Calls

Many assumptions for the disposition of blocked calls (which are also referred to as *lost calls*) have been proposed, of which the three common cases are:

- If an idle server is not immediately available, the call is cleared from the system and the source becomes idle. This is commonly called the *blocked calls cleared assumption.*

- If an idle server is not immediately available, the call is held for an interval equal to its holding time, and then the source becomes idle. If an idle server becomes available during the waiting period, it will be seized and held for an interval equal to the remaining portion of its mean holding time. This is commonly called the *blocked calls held assumption.*

- If an idle server is not immediately available, the call is queued until an idle server is available. When an idle server becomes available, it will be seized to serve the next call in queue and held for the full call-holding time. This is commonly called the *blocked calls delayed assumption.*

1.3.4 Holding-Time Distributions

A negative-exponential curve usually provides a reasonable fit for the variation in holding times encountered with nondelayed traffic-handling resources. Substituting a constant holding time equal to the average of varying holding times has a negligible effect for these applications. The effects of holding-time variations may be significant, however, when predicting the duration of delays. For example, the Crommelin-Pollaczek formulas are often used to determine service delays for resources with essentially constant holding times, such as dial-tone markers and intertoll trunks. Molnar's *Delay Probability Charts for Telephone Traffic Where the Holding Times Are Constant* graphically present data for these and similar applications.

1.4 GRADE OF SERVICE

Grade of service (GOS) is defined as the probability that offered traffic will be blocked or delayed. An absolutely nonblocking system has a GOS of zero, whereas a GOS of one indicates an absolutely blocking system. That is, the closer the grade of service is to zero, the better the system.

Every traffic problem involves three interrelated parameters: offered traffic, traffic-handling resources (servers), and service objective (grade of service). This interrelationship can be pictured as a triangle, as shown in Fig. 1-2. For a given service objective (base of triangle held constant), increasing offered traffic requires a commensurate increase in the number of servers. Similarly, decreasing the number of servers requires a corresponding decrease in the level of offered traffic.

It is important to understand that a server's GOS is a prediction of the probability of congestion (i.e., a call is blocked or delayed) at a given level of

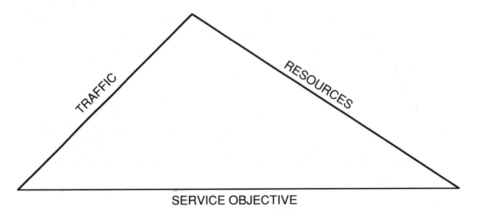

Figure 1-2.   Grade of Service Concept Diagram

offered traffic, not an absolute value. That is, a trunk-group grade of service of 0.01 does not mean that exactly one call in a hundred will be blocked during the busy hour. Rather, it means that, given a large volume of traffic, the probability of congestion will tend toward one in a hundred.

Table 1-4 lists typical grade of service specifications for traffic system design. Matching loss as used in this table refers to congestion (blocking) in a switching matrix such that input and output terminations cannot be interconnected via the interstage links. Switching matrix matching loss is not covered in this handbook but the author's *Voice Teletraffic Systems Engineering* contains an entire chapter on the subject.

Table 1-4. Typical Grade of Service Specifications

Parameter	Specification
Trunk group loss probability	0.010
Intraoffice line-to-line loss probability	0.020
Line-to-trunk outgoing matching loss probability	0.010
Trunk-to-line incoming matching loss probability	0.020
Trunk-to-trunk tandem matching loss probability	0.005
Probability dial tone delay exceeds 3 seconds	0.015
Probability operator answer delay exceeds 10 seconds	0.050

The traffic formulas found in this handbook, used to predict grades of service, are all based on probability distributions. Probability distributions are bounded by the values zero and one; therefore, a grade of service (probability of congestion) cannot be negative nor can it exceed unity. Because of this property, the probability of a call not experiencing congestion is one minus the probability of congestion, and vice versa. These relationships are expressed in Equations 1.6 and 1.7.

$$P = 1 - Q \qquad (1.6)$$

$$Q = 1 - P \qquad (1.7)$$

where P = Probability of congestion
$\quad\quad\;\; Q$ = Probability of no congestion

1.5 TRAFFIC FORMULAS AND TABLES

Table 1-5 is a selection guide for the traffic formulas contained in this handbook as a function of their typical applications. The Poisson, Erlang B, and Erlang C formulas, based on the assumption of infinite sources, are referred to as the major traffic formulas. The Binomial and Engset formulas, based on the assumption of finite sources, are used in lieu of the major traffic formulas when the number of sources is small. Figure 1-3 is a decision tree to facilitate traffic formula selection on the basis of the standard traffic assumptions.

Table 1-5. Traffic Formula Selection Guide

Typical Applications	Number of Sources	Blocked-Call Disposition	Holding-Time Distribution	Traffic Formula
Final trunk groups in North American PSTN	Infinite	Held	Constant or exponential	Poisson
Trunk groups and other nondelayed server pools	Infinite	Cleared	Constant or exponential	Erlang B
Delayed server pools	Infinite	Delayed	Exponential	Erlang C
Small PBX or remote switch trunk groups	Finite	Held	Constant or exponential	Binomial
Small line concentrators	Finite	Cleared	Constant or exponential	Engset

Representative full-availability traffic tables, selected on the basis of common telephone industry practice, are provided for the Poisson, Erlang B, Erlang C, Binomial, and Engset distributions. Full availability refers to the assumption that every source has equal access to every server. This assumption is normally true for modern traffic systems. Some older systems, however, many of which are still in use, may be limited-availability systems. Limited-availability tables, such as those found in Siemens' *Telephone Traffic Theory Tables and Charts* and ITT Standard Electrik's *Teletraffic Engineering Manual*, can be used for those systems.

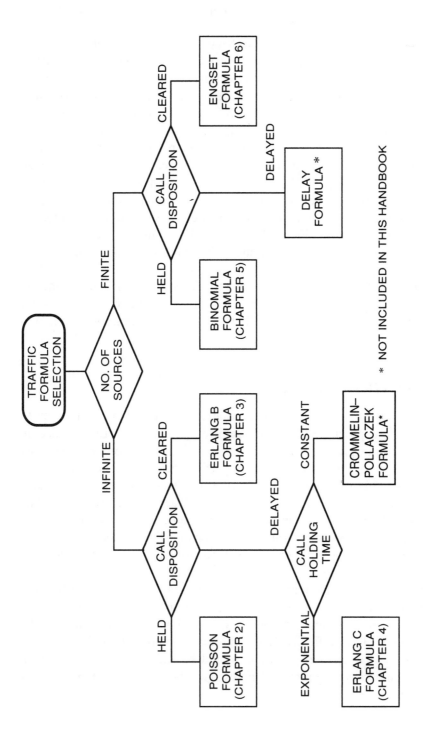

Figure 1-3   Traffic Formula Selection Decision Tree

15

Tabulated traffic data values in this handbook are rounded off to the least-significant digit as applicable to the specific table. For example, loss probability values have been rounded off to five decimal places. This level of accuracy should be more than adequate for practical applications—very low loss probabilities may indicate overdesign, which is not economically sound.

Where the parameters of a specific application do not coincide with table values, interpolation can be used. However, linear interpolation techniques are not generally satisfactory for these highly nonlinear formulas. Adequate results may be obtained with a graphic technique using semilogarithmic (semilog) graph paper, where loss probability is plotted logarithmically along the ordinate (vertical axis), and offered traffic is plotted linearly along the abscissa (horizontal axis). Figure 1-4 (page 17), a comparison of typical loss probabilities for the Poisson and Erlang B distributions, is an example of the graphic technique. Among other things it shows that, for a given loss probability, less traffic can be offered to a trunk group dimensioned using the Poisson distribution than to one containing the same number of trunks but dimensioned using the Erlang B distribution. That is, the Poisson distribution results in a more conservative design.

1.6 COMPUTER PROGRAMS

Computer programs, useful for interpolating between table values or to determine more precise values for specific applications, are provided in subsequent chapters for the Poisson, Erlang B, Erlang C, Binomial, and Engset formulas. These programs are written in BASIC because it is an easy-to-learn language and is highly standardized. It is the universal programming language for the personal computers found in homes as well as engineering offices. The programs are formatted in an interactive (i.e., dialogue) style to facilitate the user's entry of traffic parameters and include separate lines of code for each step in an attempt to make them more easily understood by those with little or no programming experience.

Readers adept at computer programming may prefer to rewrite these traffic programs, combining a number of steps into a single line of code. Alternatively, the programs can be converted to a language such as FORTRAN, which was specifically designed for computational problem solving. In any case, newly entered programs should be validated by running them against benchmarks, such as the examples in this book, before relying on their output data.

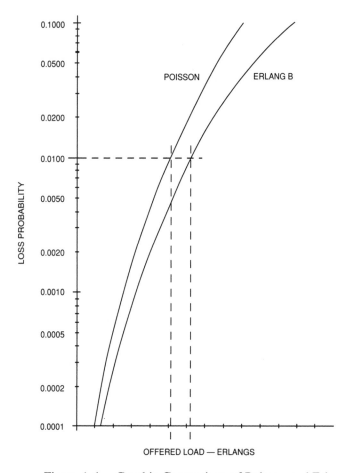

Figure 1-4.   Graphic Comparison of Poisson and Erlang B Distributions

A word of caution—computers are subject to overflow when dealing with very large numbers. This limitation is a function of the computer hardware and software, which can only process numbers within a finite range. Overflow often occurs when calculating traffic formulas, which typically involve calculation of factorials, numbers raised to the nth power, or infinite sums. The traffic programs provided herein have been written to avoid overflow conditions where possible. Overflow may still occur, however, when calculating the loss probability for a high traffic volume offered to a large number of servers, or some combination of these or other traffic parameters is encountered.

2

Poisson Distribution

The Poisson distribution is used in the North American PSTN for dimensioning final trunk groups, even though blocked calls are cleared from the system and not held. This practice is justified by extensive studies that indicate that the Poisson distribution more nearly matches actual PSTN experience than the Erlang B distribution used for foreign and military networks. It is also a more conservative distribution; that is, traffic capacity is less for a given number of trunks and loss probability (see Fig. 1-4). The Poisson distribution is based on the following assumptions:

- Calls are served in random order.
- There are an infinite number of sources.
- Blocked calls are held.
- Holding time is exponential or constant.

2.1 POISSON FORMULA

The Poisson Formula, also known as *Poisson's Exponential Binomial Limit*, is given in Equation 2.1. Equation 2.2, based on the property of probability distributions that $P = 1 - Q$, is an equivalent form of Equation 2.1, which obviates the need to sum terms to infinity.

$$P = \varepsilon^{-A} \sum_{i=N}^{\infty} \frac{A^i}{i!} \qquad (2.1)$$

$$P = 1 - \varepsilon^{-A} \sum_{i=0}^{N-1} \frac{A^i}{i!} \qquad (2.2)$$

where P = Poisson loss probability
$\qquad N$ = Number of trunks in full-availability group
$\qquad A$ = Traffic offered to group in Erlangs
$\qquad \varepsilon$ = Natural logarithm base (2.71828...)

2.2 POISSON COMPUTER PROGRAM

The following computer program can be used to calculate Equation 2.2 to determine Poisson loss probabilities. Required inputs are the number of trunks in the trunk group and the traffic offered to the group expressed in Erlangs.

```
100 REM POISSON LOSS PROBABILITY CALCULATION
110 INPUT "ENTER NUMBER OF SERVERS (N)";N
120 INPUT "ENTER OFFERED TRAFFIC IN ERLANGS (A)";A
130 LET X=1
140 LET Y=1
150 FOR I=1 TO (N-1)
160 LET X=X*A/I
170 LET Y=X+Y
180 NEXT I
190 LET Q=Y*EXP(-A)
200 PRINT USING "P = #.#####";1-Q
210 END
```

2.3 POISSON TRAFFIC CAPACITY TABLES

Poisson traffic capacity tables are used to determine the maximum amount of traffic that can be offered to a group of N trunks such that the specified loss probability (grade of service) will not be exceeded. Tables 2.1 (pages 22-26) and 2.2 (pages 27-31) present Poisson traffic capacities for 1 to 200 trunks with typical loss probabilities ranging from 0.001 to 0.1. These Poisson loss probabilities are commonly written as P.001 to P.1.

Example 2-1

Determine the traffic capacity in Erlangs and CCS for a 24-channel final trunk group such that the loss probability will not exceed 2 percent (P.02).

In Table 2-1, select the N row for 24 trunks and the P column for .02 and read 15.1 Erlangs at the intersection.

In Table 2-2, select the N row for 24 trunks and the P column for .02 and read 543.6 CCS at the intersection.

Example 2-2

Determine the number of trunks required in a final trunk group to handle 720 CCS of offered traffic at a grade of service of P.01.

In Table 2-2, select the P.01 column and read down until 730.8 CCS is found. Read across that N row to determine that 32 trunks are required.

Example 2-3

Determine the grade of service for a 48-channel final trunk group with offered traffic of 0.75 Erlangs per channel.

$$\text{Total offered traffic} = (48 \text{ chan})(0.75 \text{ Erl/chan}) = 36 \text{ Erlangs}$$

In Table 2-1, select the N row for 48 trunks and read across until 37.2 Erlangs is found. Read up that P column to determine that the grade of service is P.05.

Table 2-1. Poisson Traffic Capacity in Erlangs

No. of Trunks (N)	Traffic (A) in Erlangs for P =						
	.001	.002	.005	.010	.020	.050	.100
1	.001	.002	.005	.011	.021	.053	.106
2	.044	.065	.104	.150	.214	.358	.531
3	.192	.244	.338	.436	.567	.817	1.10
4	.428	.519	.673	.822	1.02	1.37	1.75
5	.739	.868	1.08	1.28	1.55	1.97	2.44
6	1.11	1.27	1.54	1.79	2.11	2.61	3.14
7	1.52	1.72	2.04	2.33	2.69	3.28	3.89
8	1.97	2.21	2.57	2.91	3.31	3.97	4.67
9	2.45	2.72	3.13	3.50	3.94	4.69	5.42
10	2.97	3.26	3.72	4.14	4.61	5.42	6.22
11	3.50	3.82	4.32	4.78	5.31	6.17	7.03
12	4.03	4.40	4.94	5.43	6.00	6.92	7.83
13	4.61	5.00	5.58	6.11	6.69	7.69	8.64
14	5.19	5.61	6.23	6.78	7.42	8.47	9.47
15	5.78	6.23	6.89	7.47	8.14	9.25	10.3
16	6.42	6.87	7.57	8.18	8.89	10.1	11.1
17	7.03	7.52	8.25	8.89	9.64	10.8	12.0
18	7.67	8.17	8.94	9.61	10.4	11.6	12.8
19	8.31	8.84	9.65	10.4	11.1	12.4	13.7
20	8.97	9.52	10.4	11.1	11.9	13.3	14.5
21	9.61	10.2	11.1	11.8	12.7	14.1	15.4
22	10.3	10.9	11.8	12.6	13.5	14.9	16.3
23	11.0	11.6	12.5	13.3	14.3	15.7	17.1
24	11.6	12.3	13.3	14.1	15.1	16.6	18.0
25	12.3	13.0	14.0	14.9	15.9	17.4	18.8
26	13.0	13.7	14.7	15.6	16.6	18.2	19.7
27	13.8	14.4	15.5	16.4	17.4	19.1	20.6
28	14.4	15.2	16.3	17.2	18.2	19.9	21.5
29	15.1	15.9	17.0	18.0	19.0	20.8	22.4
30	15.9	16.6	17.8	18.8	19.9	21.6	23.2
31	16.6	17.4	18.5	19.5	20.7	22.5	24.1
32	17.3	18.1	19.3	20.3	21.5	23.3	25.0
33	18.1	18.9	20.1	21.1	22.3	24.2	25.9
34	18.8	19.6	20.9	21.9	23.1	25.1	26.8
35	19.5	20.4	21.6	22.7	23.9	25.9	27.7
36	20.3	21.1	22.4	23.5	24.8	26.8	28.6
37	21.0	21.9	23.2	24.3	25.6	27.6	29.4
38	21.8	22.7	24.0	25.1	26.4	28.5	30.3
39	22.5	23.4	24.8	26.0	27.3	29.4	31.3
40	23.3	24.2	25.6	26.8	28.1	30.2	32.1

(*table continues*)

Table 2-1. Poisson Traffic Capacity in Erlangs (*Continued*)

No. of Trunks (N)	Traffic (A) in Erlangs for P =						
	.001	.002	.005	.010	.020	.050	.100
41	24.0	25.0	26.4	27.6	28.9	31.1	33.1
42	24.8	25.8	27.2	28.4	29.8	32.0	33.9
43	25.5	26.5	28.0	29.2	30.6	32.9	34.9
44	26.3	27.3	28.8	30.1	31.5	33.7	35.8
45	27.1	28.1	29.6	30.9	32.3	34.6	36.7
46	27.9	28.9	30.4	31.7	33.2	35.5	37.6
47	28.6	29.7	31.2	32.5	34.0	36.4	38.5
48	29.4	30.5	32.0	33.4	34.9	37.2	39.4
49	30.2	31.3	32.9	34.2	35.7	38.1	40.3
50	31.0	32.1	33.7	35.0	36.6	39.0	41.2
51	31.8	32.9	34.5	35.9	37.4	39.9	42.1
52	32.5	33.7	35.3	36.7	38.3	40.8	43.0
53	33.3	34.5	36.1	37.6	39.2	41.6	43.9
54	34.1	35.3	37.0	38.4	40.0	42.5	44.8
55	34.9	36.1	37.8	39.2	40.9	43.4	45.7
56	35.7	36.9	38.6	40.1	41.8	44.3	46.6
57	36.5	37.7	39.3	40.9	42.6	45.2	47.6
58	37.3	38.5	40.3	41.8	43.5	46.1	48.5
59	38.1	39.3	41.1	42.6	44.3	47.0	49.4
60	38.9	40.1	41.9	43.5	45.2	47.9	50.3
61	39.7	40.9	42.8	44.3	46.1	48.8	51.2
62	40.5	41.8	43.6	45.2	46.9	49.6	52.1
63	41.3	42.6	44.4	46.0	47.8	50.5	53.1
64	42.1	43.4	45.3	46.9	48.7	51.4	54.0
65	42.9	44.2	46.1	47.7	49.6	52.3	54.9
66	43.7	45.0	46.9	48.6	50.4	53.2	55.8
67	44.5	45.9	47.8	49.4	51.3	54.1	56.7
68	45.3	46.7	48.7	50.3	52.2	55.0	57.7
69	46.1	47.5	49.5	51.2	53.1	55.9	58.6
70	47.0	48.4	50.3	52.0	53.9	56.8	59.5
71	47.8	49.2	51.2	52.9	54.8	57.7	60.4
72	48.6	50.0	52.0	53.8	55.7	58.6	61.4
73	49.4	50.8	52.9	54.6	56.6	59.5	62.3
74	50.3	51.7	53.7	55.5	57.4	60.4	63.2
75	51.1	52.5	54.6	56.3	58.3	61.3	64.1
76	51.9	53.4	55.4	57.2	59.2	62.3	65.1
77	52.7	54.2	56.3	58.1	60.1	63.2	66.0
78	53.5	55.0	57.1	58.9	60.9	64.1	66.9
79	54.4	55.9	58.0	59.8	61.8	65.0	67.9
80	55.2	56.7	58.9	60.7	62.7	65.9	68.9

(*table continues*)

Table 2-1. Poisson Traffic Capacity in Erlangs (*Continued*)

No. of Trunks (N)	Traffic (A) in Erlangs for P=						
	.001	.002	.005	.010	.020	.050	.100
81	56.0	57.5	59.7	61.5	63.6	66.8	69.7
82	56.8	58.4	60.6	62.4	64.5	67.7	70.6
83	57.7	59.2	61.4	63.3	65.4	68.6	71.6
84	58.5	60.1	62.3	64.2	66.3	69.5	72.5
85	59.3	60.9	63.1	65.0	67.2	70.4	73.4
86	60.2	61.8	64.0	65.9	68.1	71.4	74.4
87	61.0	62.6	64.9	66.8	68.9	72.3	75.3
88	61.8	63.5	65.7	67.7	69.8	73.2	76.3
89	62.7	64.3	66.6	68.5	70.7	74.1	77.2
90	63.5	65.2	67.5	69.4	71.6	75.0	78.1
91	64.4	66.0	68.3	70.3	72.5	75.9	79.1
92	65.2	66.9	69.2	71.2	73.4	76.8	80.0
93	66.0	67.7	70.2	72.1	74.3	77.7	80.9
94	66.9	68.6	70.9	72.9	75.2	78.6	81.9
95	67.7	69.4	71.8	73.8	76.1	79.6	82.8
96	68.6	70.3	72.7	74.7	77.0	80.5	83.7
97	69.4	71.1	73.5	75.6	77.9	81.4	84.7
98	70.2	72.0	74.4	76.4	78.8	82.3	85.6
99	71.1	72.8	75.3	77.3	79.7	83.2	86.6
100	71.9	73.7	76.2	78.2	80.6	84.1	87.5
101	72.8	74.6	77.0	79.1	81.4	85.1	88.3
102	73.6	75.4	77.9	80.0	82.4	86.0	89.3
103	74.5	76.3	78.8	80.8	83.2	86.9	90.2
104	75.3	77.1	79.6	81.7	84.1	87.8	91.2
105	76.2	78.0	80.5	82.6	85.0	88.7	92.1
106	77.0	78.9	81.4	83.5	85.9	89.6	93.0
107	77.9	79.7	82.3	84.4	86.8	90.6	94.0
108	78.7	80.6	83.2	85.3	87.7	91.5	94.9
109	79.6	81.4	84.0	86.2	88.6	92.4	95.8
110	80.4	82.3	84.9	87.1	89.5	93.3	96.8
111	81.3	83.2	85.8	87.9	90.4	94.3	97.7
112	82.1	84.0	86.7	88.8	91.3	95.2	98.7
113	83.0	84.9	87.5	89.7	92.3	96.1	99.6
114	83.8	85.8	88.4	90.6	93.1	97.0	100.5
115	84.7	86.6	89.3	91.5	94.1	97.9	101.5
116	85.6	87.5	90.2	92.4	94.9	98.9	102.4
117	86.4	88.4	91.1	93.3	95.9	99.8	103.4
118	87.3	89.3	92.0	94.2	96.8	100.7	104.3
119	88.1	90.1	92.8	95.1	97.7	101.6	105.3
120	89.0	91.0	93.7	96.0	98.6	102.6	106.2

(*table continues*)

Table 2-1. Poisson Traffic Capacity in Erlangs (*Continued*)

No. of Trunks (N)	Traffic (A) in Erlangs for P =						
	.001	.002	.005	.010	.020	.050	.100
121	89.8	91.9	94.6	96.9	99.5	103.5	107.1
122	90.7	92.7	95.5	97.8	100.4	104.4	108.1
123	91.6	93.6	96.4	98.7	101.3	105.3	109.0
124	92.4	94.5	97.3	99.6	102.2	106.3	109.9
125	93.3	95.4	98.1	100.4	103.1	107.2	110.9
126	94.1	96.2	99.0	101.3	104.0	108.1	111.8
127	95.0	97.1	99.9	102.3	104.9	109.0	112.8
128	95.9	98.0	100.8	103.1	105.8	110.0	113.7
129	96.7	98.9	101.7	104.1	106.8	110.9	114.7
130	97.6	99.7	102.6	104.9	107.6	111.8	115.6
131	98.4	100.6	103.5	105.8	108.6	112.8	116.6
132	99.3	101.5	104.3	106.8	109.5	113.7	117.5
133	100.2	102.4	105.2	107.6	110.4	114.6	118.4
134	101.1	103.2	106.1	108.5	111.3	115.5	119.4
135	101.9	104.1	107.0	109.4	112.2	116.4	120.3
136	102.8	105.0	107.9	110.3	113.1	117.4	121.3
137	103.7	105.9	108.8	111.2	114.0	118.3	122.2
138	104.5	106.8	109.7	112.1	114.9	119.3	123.2
139	105.4	107.6	110.6	113.1	115.9	120.2	124.1
140	106.3	108.5	111.5	113.9	116.8	121.1	125.1
141	107.1	109.4	112.4	114.8	117.7	122.1	126.0
142	108.0	110.3	113.3	115.8	118.6	123.0	126.9
143	108.9	111.2	114.2	116.6	119.5	123.9	127.9
144	109.8	112.0	115.0	117.5	120.4	124.8	128.8
145	110.6	112.9	115.9	118.4	121.3	125.8	129.9
146	111.5	113.8	116.8	119.4	122.3	126.7	130.7
147	112.4	114.7	117.7	120.3	123.2	127.6	131.7
148	113.2	115.6	118.6	121.2	124.1	128.6	132.6
149	114.1	116.5	119.5	122.1	125.0	129.5	133.6
150	114.9	117.4	120.4	123.0	125.9	130.4	134.5
151	115.9	118.2	121.3	123.9	126.8	131.4	135.5
152	116.7	119.1	122.2	124.8	127.8	132.3	136.4
153	117.6	120.0	123.1	125.7	128.7	133.2	137.4
154	118.5	120.9	124.0	126.6	129.6	134.2	138.3
155	119.4	121.8	124.9	127.5	130.5	135.1	139.3
156	120.2	122.7	125.8	128.4	131.4	136.0	140.2
157	121.1	123.6	126.7	129.3	132.3	137.0	141.2
158	122.0	124.5	127.6	130.2	133.3	137.9	142.1
159	122.9	125.3	128.5	131.1	134.2	138.8	143.1
160	123.8	126.2	129.4	132.1	135.1	139.8	144.0

(table continues)

Table 2-1. Poisson Traffic Capacity in Erlangs (*Continued*)

No. of Trunks (N)	Traffic (A) in Erlangs for P =						
	.001	.002	.005	.010	.020	.050	.100
161	124.6	127.1	130.3	132.9	136.0	140.1	145.0
162	125.5	128.0	131.2	133.9	136.9	141.6	145.9
163	126.4	128.9	132.1	134.8	137.9	142.6	146.9
164	127.3	129.8	133.0	135.7	138.8	143.5	147.8
165	128.1	130.7	133.9	136.6	139.7	144.4	148.8
166	129.0	131.6	134.8	137.5	140.6	145.4	149.7
167	129.9	132.5	135.7	138.4	141.5	146.3	150.7
168	130.8	133.4	136.6	139.3	142.4	147.3	151.6
169	131.7	134.3	137.5	140.2	143.4	148.2	152.6
170	132.3	135.2	138.4	141.1	144.3	149.1	153.5
171	133.4	136.1	139.3	142.1	145.2	150.1	154.5
172	134.3	137.0	140.2	142.9	146.1	151.0	155.4
173	135.2	137.9	141.1	143.9	147.1	151.9	156.4
174	136.1	138.8	142.1	144.8	148.0	152.9	157.3
175	136.9	139.7	143.0	145.7	148.9	153.8	158.3
176	137.8	140.6	144.0	146.6	149.8	154.8	159.2
177	138.7	141.5	144.9	147.5	150.8	155.7	160.2
178	139.6	142.3	145.8	148.4	151.7	156.6	161.1
179	140.5	143.2	146.7	149.3	152.6	157.6	162.1
180	141.4	144.1	147.6	150.3	153.5	158.5	163.0
181	142.3	145.0	148.5	151.2	154.4	159.4	164.0
182	143.1	145.8	149.4	152.1	155.4	160.4	164.9
183	144.0	146.7	150.3	153.0	156.3	161.3	165.9
184	144.9	147.6	151.3	153.9	157.2	162.3	166.8
185	145.8	148.5	152.2	154.8	158.1	163.2	167.8
186	146.7	149.4	153.1	155.7	159.1	164.1	168.8
187	147.6	150.3	154.0	156.6	160.0	165.1	169.7
188	148.5	151.2	154.9	157.6	160.9	166.0	170.6
189	149.4	152.1	155.8	158.5	161.8	167.0	171.6
190	150.3	153.0	156.7	159.4	162.8	167.9	172.6
191	151.1	153.9	157.6	160.3	163.7	168.8	173.5
192	152.0	154.8	158.5	161.2	164.6	169.8	174.5
193	152.9	155.7	159.4	162.1	165.5	170.7	175.4
194	153.8	156.6	160.3	163.1	166.5	171.7	176.4
195	154.7	157.5	161.2	164.0	167.4	172.6	177.3
196	155.6	158.4	162.1	164.9	168.3	173.5	178.3
197	156.5	159.3	163.0	165.8	169.3	174.5	179.2
198	157.4	160.2	163.9	166.7	170.2	175.4	180.2
199	158.3	161.1	164.8	167.6	171.1	176.4	181.1
200	159.1	162.0	165.7	168.6	172.0	177.3	182.1

Table 2-2. Poisson Traffic Capacity in CCS

No. of Trunks (N)	Traffic (A) in CCS for P =						
	.001	.002	.005	.010	.020	.050	.100
1	.036	.072	.180	.396	.756	1.91	3.82
2	1.58	2.34	3.74	5.40	7.70	12.9	19.1
3	6.91	8.78	12.2	15.7	20.4	29.4	39.6
4	15.4	18.7	24.2	29.6	36.7	49.3	63.0
5	26.6	31.2	38.9	46.1	55.8	70.9	87.8
6	40.0	45.7	55.4	64.4	76.0	94.0	113.0
7	54.7	61.9	73.4	83.9	96.8	118.1	140.0
8	70.9	79.6	92.5	104.8	119.2	142.9	168.1
9	88.2	97.9	112.7	126.0	141.8	168.8	195.1
10	106.9	117.4	133.9	149.0	166.0	195.1	223.9
11	126.0	137.5	155.5	172.1	191.2	222.1	253.1
12	145.1	158.4	177.8	195.5	216.0	249.1	281.9
13	166.0	180.0	200.9	220.0	240.8	276.8	311.0
14	186.8	202.0	224.3	244.1	267.1	304.9	340.9
15	208.1	224.3	248.0	268.9	293.0	333.0	370.8
16	231.1	247.3	272.5	294.5	320.0	363.6	399.6
17	253.1	270.7	297.0	320.0	347.0	388.9	432.0
18	276.1	294.1	321.8	346.0	374.4	417.6	460.8
19	299.2	318.2	347.4	374.4	399.6	446.4	493.2
20	322.9	342.7	374.4	399.6	428.4	478.8	522.0
21	346.0	367.2	399.6	424.8	457.2	507.6	554.4
22	370.8	392.4	424.8	453.6	486.0	536.4	586.8
23	396.0	417.6	450.0	478.8	514.8	565.2	615.6
24	417.6	442.8	478.8	507.6	543.6	597.6	648.0
25	442.8	468.0	504.0	536.4	572.4	626.4	676.8
26	468.0	493.2	529.2	561.6	597.6	655.2	709.2
27	496.8	518.4	558.0	590.4	626.4	687.6	741.6
28	518.4	547.2	586.8	619.2	655.2	716.4	774.0
29	543.6	572.4	612.0	648.0	684.0	748.8	806.4
30	572.4	597.6	640.8	676.8	716.4	777.6	835.2
31	597.6	626.4	666.0	702.0	745.2	810.0	867.6
32	622.8	651.6	695.8	730.8	774.0	838.8	900.0
33	651.6	680.4	723.6	759.6	802.8	871.2	932.4
34	676.8	705.6	752.4	788.4	831.6	903.6	964.8
35	702.0	734.4	777.6	817.2	860.4	932.4	997.2
36	730.8	759.6	806.4	846.0	892.8	964.8	1029.6
37	756.0	788.4	835.2	874.8	921.6	993.6	1058.4
38	784.8	817.2	864.0	903.6	950.4	1026.0	1090.8
39	810.0	842.4	892.8	936.0	982.8	1058.4	1126.8
40	838.8	871.2	921.6	964.8	1011.6	1087.2	1156.6

(table continues)

Table 2-2. Poisson Traffic Capacity in CCS (*Continued*)

No. of Trunks (*N*)	Traffic (*A*) in CCS for *P* =						
	.001	.002	.005	.010	.020	.050	.100
41	864	900	950	994	1040	1120	1192
42	893	929	979	1022	1073	1152	1220
43	918	954	1008	1051	1102	1184	1256
44	947	983	1037	1084	1134	1213	1289
45	976	1012	1066	1112	1163	1246	1321
46	1004	1040	1094	1141	1195	1278	1354
47	1030	1069	1123	1170	1224	1310	1386
48	1058	1098	1152	1202	1256	1339	1418
49	1087	1127	1184	1231	1285	1372	1451
50	1116	1156	1213	1260	1318	1404	1483
51	1145	1184	1242	1292	1346	1436	1516
52	1170	1213	1271	1321	1379	1469	1548
53	1199	1242	1300	1354	1411	1498	1580
54	1228	1271	1332	1382	1440	1530	1613
55	1256	1300	1361	1411	1472	1562	1645
56	1285	1328	1390	1444	1505	1595	1678
57	1314	1357	1415	1472	1534	1627	1714
58	1343	1386	1451	1505	1566	1660	1746
59	1372	1415	1480	1534	1595	1692	1778
60	1400	1444	1508	1566	1627	1724	1811
61	1429	1472	1541	1595	1660	1757	1843
62	1458	1505	1570	1627	1688	1786	1876
63	1487	1534	1598	1656	1721	1818	1912
64	1516	1562	1631	1688	1753	1850	1944
65	1544	1591	1660	1717	1786	1823	1976
66	1573	1620	1688	1750	1814	1915	2009
67	1602	1652	1721	1778	1847	1948	2041
68	1631	1681	1753	1811	1879	1980	2077
69	1660	1710	1782	1843	1912	2012	2110
70	1692	1742	1811	1872	1940	2045	2142
71	1721	1771	1843	1904	1973	2077	2174
72	1750	1800	1872	1937	2005	2110	2210
73	1778	1829	1904	1966	2038	2142	2243
74	1810	1861	1933	1998	2066	2174	2275
75	1840	1890	1966	2027	2099	2207	2308
76	1868	1922	1994	2059	2131	2243	2344
77	1897	1951	2027	2092	2164	2275	2376
78	1926	1980	2056	2120	2192	2308	2408
79	1958	2012	2088	2153	2225	2340	2444
80	1987	2041	2120	2185	2257	2372	2480

(*table continues*)

Table 2-2. Poisson Traffic Capacity in CCS (*Continued*)

No. of Trunks (*N*)	Traffic (*A*) in CCS for *P* =						
	.001	.002	.005	.010	.020	.050	.100
81	2016	2070	2149	2214	2290	2405	2509
82	2045	2102	2182	2246	2322	2437	2542
83	2077	2131	2210	2279	2354	2470	2578
84	2106	2164	2243	2311	2387	2502	2610
85	2135	2192	2272	2340	2419	2534	2642
86	2167	2225	2304	2372	2452	2570	2678
87	2196	2254	2336	2405	2480	2603	2711
88	2225	2286	2365	2437	2513	2635	2747
89	2257	2315	2398	2466	2545	2668	2779
90	2286	2347	2430	2498	2578	2700	2812
91	2318	2376	2459	2531	2610	2732	2848
92	2347	2408	2491	2563	2642	2765	2880
93	2376	2437	2527	2596	2675	2797	2912
94	2408	2470	2552	2624	2707	2830	2948
95	2437	2498	2585	2657	2740	2866	2981
96	2470	2531	2617	2689	2772	2898	3013
97	2498	2560	2646	2722	2804	2930	3049
98	2527	2592	2678	2750	2837	2963	3082
99	2560	2621	2711	2783	2869	2995	3118
100	2588	2653	2743	2815	2902	3028	3150
101	2621	2686	2772	2848	2930	3064	3179
102	2650	2714	2804	2880	2966	3096	3215
103	2682	2747	2837	2909	2995	3128	3247
104	2711	2776	2866	2941	3028	3161	3283
105	2743	2808	2898	2974	3060	3193	3316
106	2772	2840	2930	3006	3092	3226	3348
107	2804	2869	2963	3038	3125	3262	3384
108	2833	2902	2995	3071	3157	3294	3416
109	2866	2930	3024	3103	3190	3326	3449
110	2894	2963	3056	3136	3222	3359	3485
111	2927	2995	3089	3164	3254	3395	3517
112	2956	3024	3121	3197	3287	3427	3553
113	2988	3056	3150	3229	3323	3460	3586
114	3017	3089	3182	3262	3352	3492	3618
115	3049	3118	3215	3294	3388	3524	3654
116	3082	3150	3247	3326	3416	3560	3686
117	3110	3182	3280	3359	3452	3593	3722
118	3143	3215	3312	3391	3485	3625	3755
119	3172	3244	3341	3424	3517	3658	3791
120	3204	3276	3373	3456	3550	3694	3823

(*table continues*)

Table 2-2. Poisson Traffic Capacity in CCS (*Continued*)

No. of Trunks (N)	Traffic (A) in CCS for P =						
	.001	.002	.005	.010	.020	.050	.100
121	3233	3308	3406	3488	3582	3726	3856
122	3265	3337	3438	3521	3614	3758	3892
123	3298	3370	3470	3553	3647	3791	3924
124	3326	3402	3503	3586	3679	3827	3956
125	3359	3434	3532	3614	3712	3859	3992
126	3388	3463	3564	3647	3744	3892	4025
127	3420	3496	3596	3683	3776	3924	4061
128	3452	3528	3629	3712	3809	3960	4093
129	3481	3560	3661	3748	3845	3992	4129
130	3514	3589	3694	3776	3874	4025	4162
131	3542	3622	3726	3809	3910	4061	4198
132	3575	3654	3755	3845	3942	4093	4230
133	3607	3686	3787	3874	3974	4126	4262
134	3640	3715	3820	3906	4007	4158	4298
135	3668	3748	3852	3938	4039	4190	4331
136	3701	3780	3884	3971	4072	4226	4367
137	3733	3812	3917	4003	4104	4259	4399
138	3762	3845	3949	4036	4136	4295	4435
139	3794	3874	3982	4072	4172	4327	4468
140	3827	3906	4014	4100	4205	4360	4504
141	3856	3938	4046	4133	4237	4396	4536
142	3888	3971	4079	4169	4270	4428	4568
143	3920	4003	4111	4198	4302	4460	4604
144	3953	4032	4140	4230	4334	4493	4637
145	3982	4064	4172	4262	4367	4529	4676
146	4014	4097	4205	4298	4403	4561	4705
147	4046	4129	4237	4331	4435	4594	4741
148	4075	4162	4270	4363	4468	4630	4774
149	4108	4194	4302	4396	4500	4662	4810
150	4136	4226	4334	4428	4532	4694	4842
151	4172	4255	4367	4460	4565	4730	4878
152	4201	4288	4399	4493	4601	4763	4910
153	4234	4320	4432	4525	4633	4795	4946
154	4266	4352	4464	4558	4666	4831	4979
155	4298	4385	4496	4590	4698	4864	5015
156	4327	4417	4529	4622	4730	4896	5047
157	4360	4450	4561	4655	4763	4932	5083
158	4392	4482	4594	4687	4799	4964	5116
159	4424	4511	4626	4720	4831	4997	5152
160	4457	4543	4658	4756	4864	5033	5184

(table continues)

Table 2-2. Poisson Traffic Capacity in CCS (*Continued*)

No. of Trunks (*N*)	Traffic (*A*) in CCS for *P* =						
	.001	.002	.005	.010	.020	.050	.100
161	4486	4576	4691	4784	4896	5044	5220
162	4518	4608	4723	4820	4928	5098	5252
163	4550	4640	4756	4853	4964	5134	5288
164	4583	4673	4788	4885	4997	5166	5321
165	4612	4705	4820	4918	5029	5198	5357
166	4644	4738	4853	4950	5062	5234	5389
167	4676	4770	4885	4982	5094	5267	5425
168	4709	4802	4918	5015	5126	5303	5458
169	4741	4835	4950	5047	5162	5335	5494
170	4763	4867	4982	5080	5195	5368	5526
171	4802	4900	5015	5116	5227	5404	5562
172	4835	4932	5047	5144	5260	5436	5594
173	4867	4964	5080	5180	5296	5468	5630
174	4900	4997	5116	5213	5328	5504	5663
175	4928	5029	5148	5245	5360	5537	5699
176	4961	5062	5184	5278	5393	5573	5731
177	4993	5094	5216	5310	5429	5605	5767
178	5026	5123	5249	5342	5461	5638	5800
179	5058	5155	5281	5375	5494	5674	5836
180	5090	5188	5314	5411	5526	5706	5868
181	5123	5220	5346	5443	5558	5738	5904
182	5152	5249	5378	5476	5594	5774	5936
183	5184	5281	5411	5508	5627	5807	5972
184	5216	5314	5447	5540	5659	5843	6005
185	5249	5346	5479	5573	5692	5875	6041
186	5281	5378	5512	5605	5728	5908	6077
187	5314	5411	5544	5638	5760	5944	6109
188	5346	5443	5576	5674	5792	5976	6142
189	5378	5476	5609	5706	5825	6012	6178
190	5411	5508	5641	5738	5861	6044	6214
191	5440	5540	5674	5771	5893	6077	6246
192	5472	5573	5706	5803	5926	6113	6282
193	5504	5605	5738	5836	5958	6145	6314
194	5537	5638	5771	5872	5994	6181	6350
195	5569	5670	5803	5904	6026	6214	6383
196	5602	5702	5836	5936	6059	6246	6419
197	5634	5735	5868	5969	6095	6282	6451
198	5666	5767	5900	6001	6127	6314	6487
199	5699	5800	5933	6034	6160	6350	6520
200	5728	5832	5965	6070	6192	6383	6556

3

Erlang B Distribution

The Erlang B distribution is used in the North American PSTN for dimensioning high-usage trunk groups in alternate routing systems (blocked calls offered to another high-usage trunk group or to a final trunk group), and for all trunk group dimensioning in foreign and military networks. In addition, it is applicable to common equipment server pools which are not provided with a waiting queue (immediate service). The Erlang B distribution is based on the following assumptions:

- Calls are served in random order.
- There are an infinite number of sources.
- Blocked calls are cleared.
- Holding times are exponential or constant .

3.1 ERLANG B FORMULA

The Erlang B Formula, also known as the *Erlang formula of the first kind* [$E_1(N, A)$], is given in Equation 3.1.

$$B = E_1(N, A) = \frac{\dfrac{A^N}{N!}}{\displaystyle\sum_{i=0}^{N} \dfrac{A^i}{i!}} \qquad\qquad (3.1)$$

where B = Erlang B loss probability

  N = Number of trunks in full-availability group

  A = Traffic offered to group in Erlangs

3.2   ERLANG B COMPUTER PROGRAM

The following computer program can be used to calculate Equation 3.1 to determine Erlang B loss probabilities. Required inputs are the number of trunks in the trunk group and the traffic offered to the group expressed in Erlangs.

```
100 REM ERLANG B LOSS PROBABILITY CALCULATION
110 INPUT "ENTER NUMBER OF SERVERS (N)";N
120 INPUT "ENTER OFFERED TRAFFIC IN ERLANGS (A)";A
130 LET X=1
140 LET Y=1
150 FOR I=1 TO N
160 LET X=X*A/I
170 LET Y=X+Y
180 NEXT I
190 PRINT USING "B = #.#####";X/Y
200 END
```

3.3   ERLANG B TRAFFIC CAPACITY TABLES

Erlang B traffic capacity tables are used to determine the maximum amount of traffic that can be offered to a group of N trunks such that the specified loss probability (grade of service) will not be exceeded. Tables 3.1 (pages 36–41) and 3.2 (pages 42–47) present Erlang B traffic capacities for 1 to 200 trunks with typical loss probabilities ranging from 0.001 to 0.1. These Erlang B loss probabilities are commonly written as B.001 to B.1. The Erlang B formula yields relatively linear traffic capacity values for trunk groups with more than

200 trunks. Therefore, linear interpolation can be used between the given table values for those applications. The following examples illustrate typical table usage:

Example 3-1

Determine the traffic capacity in Erlangs and CCS for a 24-channel high-usage trunk group such that the loss probability will not exceed 2 percent (B.02).

In Table 3-1, select the *N* row for 24 trunks and the *B* column for .02 and read 16.6 Erlangs at the intersection.

In Table 3-2, select the *N* row for 24 trunks and the *P* column for .02 and read 597.6 CCS at the intersection.

Example 3-2

Determine the number of trunks required in a high-usage trunk group to handle 720 CCS of offered traffic at a grade of service of B.01.

In Table 3-2, select the B.01 column and read down until 730.8 CCS is found. Read across that *N* row to determine that 30 trunks are required.

Example 3-3

Determine the grade of service for a 48-channel high-usage trunk group with offered traffic of 0.75 Erlangs per trunk (36 Erlangs).

$$A = (48 \text{ chan}) (0.75 \text{ Erl/chan}) = 36 \text{ Erlangs}$$

In Table 3-1, select the *N* row for 48 trunks and read across until 36.1 Erlangs is found. Read up that *B* column to determine that the grade of service is B.01.

Table 3-1. Erlang B Traffic Capacity in Erlangs

No. of Trunks (N)	Traffic (A) in Erlangs for B =						
	.001	.002	.005	.010	.020	.050	.100
1	.001	.002	.005	.011	.021	.053	.111
2	.046	.066	.106	.153	.224	.382	.595
3	.194	.249	.349	.456	.603	.900	1.27
4	.440	.536	.702	.870	1.09	1.52	2.05
5	.763	.900	1.13	1.36	1.66	2.22	2.88
6	1.15	1.33	1.62	1.91	2.28	2.96	3.76
7	1.58	1.80	2.16	2.50	2.94	3.74	4.67
8	2.05	2.31	2.73	3.13	3.63	4.54	5.60
9	2.56	2.86	3.33	3.78	4.34	5.37	6.55
10	3.09	3.43	3.96	4.46	5.08	6.22	7.51
11	3.65	4.02	4.61	5.16	5.84	7.08	8.49
12	4.23	4.64	5.28	5.88	6.62	7.95	9.47
13	4.83	5.27	5.96	6.61	7.41	8.83	10.5
14	5.45	5.92	6.66	7.35	8.20	9.73	11.5
15	6.08	6.58	7.38	8.11	9.01	10.6	12.5
16	6.72	7.26	8.10	8.87	9.83	11.5	13.5
17	7.38	7.95	8.83	9.65	10.7	12.5	14.5
18	8.05	8.64	9.58	10.4	11.5	13.4	15.6
19	8.72	9.35	10.3	11.2	12.3	14.3	16.6
20	9.41	10.1	11.1	12.0	13.2	15.3	17.6
21	10.1	10.8	11.9	12.8	14.0	16.2	18.7
22	10.8	11.5	12.6	13.7	14.9	17.1	19.7
23	11.5	12.3	13.4	14.5	15.8	18.1	20.7
24	12.2	13.0	14.2	15.3	16.6	19.0	21.8
25	13.0	13.8	15.0	16.1	17.5	20.0	22.8
26	13.7	14.5	15.8	17.0	18.4	20.9	23.9
27	14.4	15.3	16.6	17.8	19.3	21.9	24.9
28	15.2	16.1	17.4	18.6	20.2	22.9	26.0
29	15.9	16.8	18.2	19.5	21.0	23.8	27.1
30	16.7	17.6	19.0	20.3	21.9	24.8	28.1
31	17.4	18.4	19.9	21.2	22.8	25.8	29.2
32	18.2	19.2	20.7	22.1	23.7	26.8	30.2
33	19.0	20.0	21.5	22.9	24.6	27.7	31.3
34	19.7	20.8	22.3	23.8	25.5	28.7	32.4
35	20.5	21.6	23.2	24.6	26.4	29.7	33.4
36	21.3	22.4	24.0	25.5	27.3	30.7	34.5
37	22.1	23.2	24.9	26.4	28.3	31.6	35.6
38	22.9	24.0	25.7	27.3	29.2	32.6	36.6
39	23.7	24.8	26.5	28.1	30.1	33.6	37.7
40	24.4	25.6	27.4	29.0	31.0	34.6	38.8

(table continues)

Table 3-1. Erlang B Traffic Capacity in Erlangs (*Continued*)

No. of Trunks (N)	.001	.002	.005	.010	.020	.050	.100
			Traffic (A) in Erlangs for B =				
41	25.2	26.4	28.2	29.9	31.9	35.6	39.9
42	26.0	27.2	29.1	30.8	32.8	36.6	40.9
43	26.8	28.1	29.9	31.7	33.8	37.6	42.0
44	27.6	28.9	30.8	32.5	34.7	38.6	43.1
45	28.5	29.7	31.7	33.4	35.6	39.6	44.2
46	29.3	30.5	32.5	34.3	36.5	40.5	45.2
47	30.1	31.4	33.4	35.2	37.5	41.5	46.3
48	30.9	32.2	34.2	36.1	38.4	42.5	47.4
49	31.7	33.0	35.1	37.0	39.3	43.5	48.5
50	32.5	33.9	36.0	37.9	40.3	44.5	49.6
51	33.3	34.7	36.9	38.8	41.2	45.5	50.6
52	34.2	35.6	37.7	39.7	42.1	46.5	51.7
53	35.0	36.4	38.6	40.6	43.1	47.5	52.8
54	35.8	37.3	39.5	41.5	44.0	48.5	53.9
55	36.6	38.1	40.4	42.4	44.9	49.5	55.0
56	37.5	38.9	41.2	43.3	45.9	50.5	56.1
57	38.3	39.8	42.1	44.2	46.8	51.5	57.1
58	39.1	40.6	43.0	45.1	47.8	52.6	58.2
59	40.0	41.5	43.9	46.0	48.7	53.6	59.3
60	40.8	42.4	44.8	46.9	49.6	54.6	60.4
61	41.6	43.2	45.6	47.9	50.6	55.6	61.5
62	42.5	44.1	46.5	48.8	51.5	56.6	62.6
63	43.3	44.9	47.7	49.7	52.5	57.6	63.7
64	44.2	45.8	48.3	50.6	53.4	58.6	64.8
65	45.0	46.7	49.2	51.5	54.4	59.6	65.8
66	45.8	47.5	50.1	52.4	55.3	60.6	66.9
67	46.7	48.4	51.0	53.4	56.3	61.6	68.0
68	47.5	49.2	51.9	54.3	57.2	62.6	69.1
69	48.4	50.1	52.8	55.2	58.2	63.7	70.2
70	49.2	51.0	53.7	56.1	59.1	64.7	71.3
71	50.1	51.9	54.6	57.0	60.1	65.7	72.4
72	50.9	52.7	55.5	58.0	61.0	66.7	73.5
73	51.8	53.6	56.4	58.9	62.0	67.7	74.6
74	52.7	54.5	57.3	59.8	62.9	68.7	75.6
75	53.5	55.3	58.2	60.7	63.9	69.7	76.7
76	54.4	56.2	59.1	61.7	64.9	70.8	77.8
77	55.2	57.1	60.0	62.6	65.8	71.8	78.9
78	56.1	58.0	60.9	63.5	66.8	72.8	80.0
79	57.0	58.8	61.8	64.4	67.7	73.8	81.1
80	57.8	59.7	62.7	65.4	68.7	74.8	82.2

(*table continues*)

Table 3-1. Erlang B Traffic Capacity in Erlangs (*Continued*)

No. of Trunks (*N*)	Traffic (*A*) in Erlangs for *B* =						
	.001	.002	.005	.010	.020	.050	.100
81	58.7	60.6	63.6	66.3	69.6	75.8	83.3
82	59.5	61.5	64.5	67.2	70.6	76.9	84.4
83	60.4	62.4	65.4	68.2	71.6	77.9	85.5
84	61.3	63.2	66.3	69.1	72.5	78.9	86.6
85	62.1	64.1	67.2	70.0	73.5	79.9	87.7
86	63.0	65.0	68.1	70.9	74.5	80.9	88.8
87	63.9	65.9	69.0	71.9	75.4	82.0	89.9
88	64.7	66.8	69.9	72.8	76.4	83.0	91.0
89	65.6	67.7	70.8	73.7	77.3	84.0	92.1
90	66.5	68.6	71.8	74.7	78.3	85.0	93.1
91	67.4	69.4	72.7	75.6	79.3	86.0	94.2
92	68.2	70.3	73.6	76.6	80.2	87.1	95.3
93	69.1	71.2	74.5	77.5	81.2	88.1	96.4
94	70.0	72.1	75.4	78.4	82.2	89.1	97.5
95	70.9	73.0	76.3	79.4	83.1	90.1	98.6
96	71.7	73.9	77.2	80.3	84.1	91.1	99.7
97	72.6	74.8	78.2	81.2	85.1	92.2	100.8
98	73.5	75.7	79.1	82.2	86.0	93.2	101.9
99	74.4	76.6	80.0	83.1	87.0	94.2	103.0
100	75.2	77.5	80.9	84.1	88.0	95.2	104.1
101	76.1	78.4	81.8	85.0	88.9	96.3	105.2
102	77.0	79.3	82.8	85.9	89.9	97.3	106.3
103	77.9	80.2	83.7	86.9	90.9	98.3	107.4
104	78.8	81.1	84.6	87.8	91.9	99.3	108.5
105	79.7	81.9	85.5	88.8	92.8	100.4	109.6
106	80.5	82.9	86.4	89.7	93.8	101.4	110.7
107	81.4	83.8	87.4	90.7	94.8	102.4	111.8
108	82.3	84.7	88.3	91.6	95.7	103.4	112.9
109	83.2	85.6	89.2	92.5	96.7	104.5	114.0
110	84.1	86.5	90.1	93.5	97.7	105.5	115.1
111	85.0	87.4	91.0	94.4	98.7	106.5	116.2
112	85.9	88.3	92.0	95.4	99.6	107.5	117.3
113	86.7	89.2	92.9	96.3	100.6	108.6	118.4
114	87.6	90.1	93.8	97.3	101.6	109.6	119.5
115	88.5	91.0	94.7	98.2	102.5	110.6	120.6
116	89.4	91.9	95.7	99.2	103.5	111.7	121.7
117	90.3	92.8	96.6	100.1	104.5	112.7	122.8
118	91.2	93.7	97.5	101.1	105.5	113.7	123.9
119	92.1	94.6	98.5	102.0	106.4	114.7	125.0
120	93.0	95.5	99.4	103.0	107.4	115.8	126.1

(*table continues*)

Table 3-1. Erlang B Traffic Capacity in Erlangs (*Continued*)

No. of Trunks (*N*)	Traffic (*A*) in Erlangs for *B* =						
	.001	.002	.005	.010	.020	.050	.100
121	93.8	96.4	100.3	103.9	108.4	116.8	127.2
122	94.7	97.3	101.2	104.9	109.4	117.8	128.3
123	95.6	98.2	102.2	105.8	110.3	118.9	129.4
124	96.5	99.1	103.1	106.8	111.3	119.9	130.5
125	97.4	100.0	104.0	107.7	112.3	120.9	131.6
126	98.3	100.9	105.0	108.7	113.3	122.0	132.7
127	99.2	101.8	105.9	109.6	114.2	123.0	133.8
128	100.1	102.7	106.8	110.6	115.2	124.0	134.9
129	101.0	103.7	107.8	111.5	116.2	125.1	136.0
130	101.9	104.6	108.7	112.5	117.2	126.1	137.1
131	102.8	105.5	109.6	113.4	118.2	127.1	138.2
132	103.7	106.4	110.5	114.4	119.1	128.2	139.3
133	104.6	107.3	111.5	115.3	120.1	129.2	140.4
134	105.5	108.2	112.4	116.3	121.1	130.2	141.5
135	106.4	109.1	113.3	117.2	122.1	131.3	142.6
136	107.3	110.0	114.3	118.2	123.1	132.3	143.7
137	108.2	111.0	115.2	119.1	124.0	133.3	144.8
138	109.1	111.9	116.2	120.1	125.0	134.3	145.9
139	110.0	112.8	117.1	121.0	126.0	135.4	147.0
140	110.9	113.7	118.0	122.0	127.0	136.4	148.1
141	111.8	114.6	119.0	122.9	128.0	137.4	149.2
142	112.7	115.5	119.9	123.9	128.9	138.5	150.3
143	113.6	116.4	120.8	124.8	129.9	139.5	151.4
144	114.5	117.4	121.8	125.8	130.9	140.5	152.5
145	115.4	118.3	122.7	126.7	131.9	141.6	153.6
146	116.3	119.2	123.6	127.7	132.9	142.6	154.7
147	117.2	120.1	124.6	128.6	133.8	143.6	155.8
148	118.1	121.0	125.5	129.6	134.8	144.7	156.9
149	119.0	121.9	126.5	130.6	135.8	145.7	158.0
150	119.9	122.9	127.4	131.6	136.8	146.7	159.1
151	120.8	123.8	128.3	132.5	137.8	147.8	160.3
152	121.7	124.7	129.3	133.5	138.7	148.8	161.4
153	122.6	125.6	130.2	134.5	139.7	149.8	162.5
154	123.5	126.5	131.2	135.4	140.7	150.9	163.6
155	124.4	127.5	132.1	136.4	141.7	150.9	164.7
156	125.3	128.4	133.0	137.4	142.7	152.9	165.8
157	126.2	129.3	134.0	138.3	143.6	153.9	166.9
158	127.1	130.2	134.9	139.3	144.6	154.9	168.0
159	128.0	131.1	135.9	140.3	145.6	156.0	169.1
160	129.0	132.1	136.8	141.2	146.6	157.0	170.2

(*table continues*)

Table 3-1. Erlang B Traffic Capacity in Erlangs (*Continued*)

No. of Trunks (*N*)	Traffic (*A*) in Erlangs for *B* =						
	.001	.002	.005	.010	.020	.050	.100
161	129.9	133.0	137.7	142.2	147.6	158.0	171.3
162	130.8	133.9	138.7	143.2	148.6	159.1	172.4
163	131.7	134.8	139.6	144.1	149.6	160.1	173.5
164	132.6	135.8	140.6	145.1	150.5	161.1	174.6
165	133.5	136.7	141.5	146.1	151.5	162.2	175.7
166	134.4	137.6	142.4	147.0	152.5	163.2	176.8
167	135.3	138.5	143.4	148.0	153.5	164.2	177.9
168	136.2	139.4	144.3	148.9	154.5	165.3	179.0
169	137.1	140.4	145.3	149.9	155.5	166.3	180.1
170	138.1	141.3	146.2	150.8	156.5	167.4	181.2
171	139.0	142.2	147.2	151.8	157.5	168.4	182.3
172	139.9	143.1	148.1	152.8	158.5	169.4	183.4
173	140.8	144.1	149.1	153.7	159.4	170.5	184.5
174	141.7	145.0	150.0	154.7	160.4	171.5	185.6
175	142.7	145.9	150.9	155.6	161.4	172.6	186.7
176	143.6	146.9	151.9	156.6	162.4	173.6	187.8
177	144.5	147.8	152.8	157.5	163.4	174.7	188.9
178	145.5	148.7	153.8	158.5	164.4	175.7	190.0
179	146.3	149.6	154.7	159.4	165.4	176.8	191.1
180	147.3	150.6	155.7	160.4	166.4	177.8	192.2
181	148.2	151.5	156.6	161.4	167.4	178.8	193.4
182	149.1	152.4	157.6	162.3	168.4	179.9	194.5
183	150.0	153.3	158.5	163.3	169.3	180.9	195.6
184	150.9	154.3	159.5	164.3	170.3	181.9	196.7
185	151.8	155.2	160.4	165.2	171.3	183.0	197.8
186	152.8	156.1	161.4	166.2	172.3	184.0	198.9
187	153.7	157.1	162.3	167.2	173.3	185.0	200.0
188	154.6	158.0	163.3	168.1	174.3	186.1	201.1
189	155.5	158.9	164.2	169.1	175.3	187.1	202.2
190	156.4	159.8	165.2	170.1	176.3	188.1	203.3
191	157.4	160.7	166.1	171.0	177.3	189.2	204.4
192	158.3	161.7	167.0	172.0	178.3	190.2	205.5
193	159.2	162.6	167.9	173.0	179.3	191.2	206.6
194	160.2	163.6	168.9	173.9	180.3	192.3	207.7
195	161.1	164.5	169.8	174.5	181.2	193.3	208.8
196	162.0	165.4	170.8	175.9	182.2	194.3	209.9
197	162.9	166.3	171.7	176.8	183.2	195.4	211.0
198	163.8	167.3	172.7	177.8	184.2	196.4	212.1
199	164.7	168.2	173.6	178.8	185.2	197.4	213.2
200	165.6	169.2	174.6	179.7	186.2	198.5	214.3

(*table continues*)

Table 3-1. Erlang B Traffic Capacity in Erlangs (*Continued*)

No. of Trunks (*N*)	Traffic (*A*) in Erlangs for *B* =						
	.001	.002	.005	.010	.020	.050	.100
202	167.5	171.0	176.5	181.7	188.1	200.6	216.5
204	169.3	172.9	178.4	183.6	190.1	202.7	218.7
206	171.2	174.8	180.4	185.5	192.1	204.7	221.0
208	173.0	176.6	182.3	187.5	194.1	206.8	223.2
210	174.8	178.5	184.2	189.4	196.1	208.9	225.4
212	176.7	180.4	186.1	191.4	198.1	211.0	227.6
214	178.5	182.2	188.0	193.3	200.0	213.0	229.8
216	180.4	184.1	189.9	195.2	202.0	215.1	232.0
218	182.2	186.0	191.8	197.2	204.0	217.2	234.2
220	184.1	187.8	193.7	199.1	206.0	219.3	236.4
222	185.9	189.7	195.6	201.1	208.0	221.4	238.6
224	187.8	191.6	197.5	203.0	210.0	223.4	240.9
226	189.6	193.5	199.4	204.9	212.0	225.5	243.1
228	191.5	195.3	201.3	206.9	213.9	227.6	245.3
230	193.3	197.2	203.2	208.8	215.9	229.7	247.5
232	195.2	199.1	205.1	210.8	217.9	231.8	249.7
234	197.1	201.0	207.1	212.7	219.9	233.8	251.9
236	198.9	202.8	209.0	214.7	221.9	235.9	254.1
238	200.8	204.7	210.9	216.6	223.9	238.0	256.3
240	202.6	206.6	212.8	218.6	225.9	240.1	258.6
242	204.5	208.5	214.7	220.5	227.9	242.2	260.8
244	206.3	210.4	216.6	222.5	229.9	244.3	263.0
246	208.2	212.2	218.5	224.4	231.8	246.3	265.2
248	210.1	214.1	220.4	226.3	233.9	248.4	267.4
250	211.9	216.0	222.4	228.3	235.8	250.5	269.6
300	258.6	263.2	270.4	277.1	285.7	302.6	325.0
350	305.7	310.8	318.7	326.2	335.7	354.8	380.4
400	353.0	358.5	367.2	375.3	385.9	407.1	435.8
450	400.5	406.4	415.8	424.6	436.1	459.4	491.3
500	448.2	454.5	464.5	474.0	486.4	511.8	546.7
550	496.1	502.8	513.4	523.6	536.8	564.2	602.2
600	543.9	551.0	562.3	573.1	587.2	616.5	657.7
650	592.0	599.6	611.4	622.8	637.7	669.0	713.2
700	640.1	647.9	660.4	672.4	688.2	721.4	768.7
750	688.4	696.5	709.6	721.6	738.8	773.9	824.2
800	736.6	741.5	758.7	771.8	789.3	826.4	879.7
850	785.0	790.2	808.7	821.7	840.0	878.9	935.3
900	833.3	842.5	857.2	871.5	890.6	931.4	990.8
950	881.8	891.3	906.6	921.4	941.3	983.7	1046.4
1000	930.3	940.1	955.9	971.2	991.9	1036.0	1102.0

Table 3-2. Erlang B Traffic Capacity in CCS

No. of Trunks (N)	Traffic (A) in CCS for B =						
	.001	.002	.005	.010	.020	.050	.100
1	.036	.072	.180	.396	.756	1.91	4.00
2	1.66	2.38	3.82	5.51	8.06	13.8	21.4
3	6.98	8.96	12.6	16.4	21.7	32.4	45.7
4	15.8	19.3	25.3	31.3	39.2	54.7	73.8
5	27.5	32.4	40.7	49.0	59.8	79.9	103.7
6	41.4	47.9	58.3	68.8	82.1	106.6	135.4
7	56.9	64.8	77.8	90.0	105.8	134.6	168.1
8	73.8	83.2	98.3	112.7	130.7	163.4	201.6
9	92.2	103.0	119.9	136.1	156.2	193.3	235.8
10	111.2	123.5	142.6	160.6	182.9	223.9	270.4
11	131.4	144.7	166.0	185.8	210.2	254.9	305.6
12	152.3	167.0	190.1	211.7	238.3	286.2	340.9
13	173.9	189.7	214.6	238.0	266.8	317.9	378.0
14	196.2	213.1	239.8	264.6	295.2	350.3	414.0
15	218.9	236.9	265.7	292.0	324.4	381.6	450.0
16	241.9	261.4	291.6	319.3	353.9	414.0	486.0
17	265.7	286.2	317.9	347.4	385.2	450.0	522.0
18	289.8	311.0	344.9	374.4	414.0	482.4	561.6
19	313.9	336.6	370.8	403.2	442.8	514.8	597.6
20	338.8	363.6	399.6	432.0	475.2	550.8	633.6
21	363.6	388.8	428.4	460.8	504.0	583.2	673.2
22	388.8	414.0	453.6	493.2	536.4	615.6	709.2
23	414.0	442.8	482.4	522.0	568.8	651.6	745.2
24	439.2	468.0	511.2	550.8	597.6	684.0	784.8
25	468.0	496.8	540.0	579.6	630.0	720.0	820.8
26	493.2	522.0	568.8	612.0	662.4	752.4	860.4
27	518.4	550.8	597.6	640.8	694.8	788.4	896.4
28	547.2	579.6	626.4	669.6	727.2	824.4	936.0
29	572.4	604.8	655.2	702.0	756.0	856.8	975.6
30	601.2	633.6	684.0	730.8	788.4	892.8	1011.6
31	626.4	662.4	716.4	763.2	820.8	928.8	1051.2
32	655.2	691.2	745.2	795.6	853.2	964.8	1087.2
33	684.0	720.0	774.0	824.4	885.6	997.2	1126.8
34	709.2	748.8	802.8	856.8	918.0	1033.2	1166.4
35	738.0	777.6	835.2	885.6	950.4	1069.2	1202.4
36	766.8	806.4	864.0	918.0	982.8	1105.2	1242.0
37	795.6	835.2	896.4	950.4	1018.8	1137.6	1281.6
38	824.4	864.0	925.2	982.8	1051.2	1173.6	1317.6
39	853.2	892.8	954.0	1011.6	1083.6	1209.6	1357.2
40	878.4	921.6	986.4	1044.0	1116.0	1245.6	1396.8

(table continues)

Table 3-2. Erlang B Traffic Capacity in CCS (*Continued*)

No. of Trunks (*N*)	Traffic (*A*) in CCS for *B* =						
	.001	.002	.005	.010	.020	.050	.100
41	907	950	1015	1076	1148	1282	1436
42	936	979	1047	1109	1181	1318	1472
43	965	1012	1076	1141	1217	1354	1512
44	994	1040	1109	1170	1249	1390	1552
45	1026	1069	1141	1202	1282	1426	1591
46	1055	1098	1170	1235	1314	1458	1627
47	1084	1130	1202	1267	1350	1494	1667
48	1112	1159	1231	1300	1382	1530	1706
49	1141	1188	1264	1332	1415	1566	1746
50	1170	1220	1296	1364	1451	1602	1786
51	1199	1249	1328	1397	1483	1638	1822
52	1231	1282	1357	1429	1516	1674	1861
53	1260	1310	1390	1462	1552	1710	1901
54	1289	1343	1422	1494	1584	1746	1940
55	1318	1372	1454	1526	1616	1782	1980
56	1350	1400	1483	1559	1652	1818	2020
57	1379	1433	1516	1591	1685	1854	2056
58	1408	1462	1548	1624	1721	1894	2095
59	1440	1494	1580	1656	1753	1930	2135
60	1469	1526	1613	1688	1786	1966	2174
61	1498	1555	1642	1724	1822	2002	2214
62	1530	1588	1674	1757	1854	2038	2254
63	1559	1616	1717	1789	1890	2074	2293
64	1591	1649	1739	1822	1922	2110	2333
65	1620	1681	1771	1854	1958	2146	2369
66	1649	1710	1804	1886	1991	2182	2408
67	1681	1742	1836	1922	2027	2218	2448
68	1710	1771	1868	1955	2059	2254	2488
69	1742	1804	1901	1987	2095	2293	2527
70	1771	1836	1933	2020	2128	2329	2567
71	1804	1868	1966	2052	2164	2365	2606
72	1832	1897	1998	2088	2196	2401	2646
73	1865	1930	2030	2120	2232	2437	2686
74	1897	1962	2063	2153	2264	2473	2722
75	1926	1991	2095	2185	2300	2509	2761
76	1958	2023	2128	2221	2336	2549	2801
77	1987	2056	2160	2254	2369	2585	2840
78	2020	2088	2192	2286	2405	2621	2880
79	2052	2117	2225	2318	2437	2657	2920
80	2081	2149	2257	2354	2473	2693	2959

(*table continues*)

Table 3-2. Erlang B Traffic Capacity in CCS (*Continued*)

No. of Trunks (N)	Traffic (A) in CCS for B =						
	.001	.002	.005	.010	.020	.050	.100
81	2113	2182	2290	2387	2506	2729	2999
82	2142	2214	2322	2419	2542	2768	3038
83	2174	2246	2354	2455	2578	2804	3078
84	2207	2275	2387	2488	2610	2840	3118
85	2236	2308	2419	2520	2646	2876	3157
86	2268	2340	2452	2552	2682	2912	3197
87	2300	2372	2484	2588	2714	2952	3236
88	2329	2405	2516	2621	2750	2988	3276
89	2362	2437	2549	2653	2783	3024	3316
90	2394	2470	2585	2689	2819	3060	3352
91	2426	2498	2617	2722	2855	3096	3391
92	2455	2531	2650	2758	2887	3136	3431
93	2488	2563	2682	2790	2923	3172	3470
94	2520	2596	2714	2822	2959	3208	3510
95	2552	2628	2747	2858	2992	3244	3550
96	2581	2660	2779	2891	3028	3280	3589
97	2914	2693	2815	2923	3064	3319	3629
98	2646	2725	2848	2959	3096	3355	3668
99	2678	2758	2880	2992	3132	3391	3708
100	2707	2790	2912	3028	3168	3427	3748
101	2740	2822	2945	3060	3200	3467	3787
102	2772	2855	2981	3092	3236	3503	3827
103	2804	2887	3013	3128	3272	3539	3866
104	2837	2920	3046	3161	3308	3575	3906
105	2869	2948	3078	3197	3341	3614	3946
106	2898	2984	3110	3229	3377	3650	3985
107	2930	3017	3146	3265	3413	3686	4025
108	2963	3049	3179	3298	3445	3722	4064
109	2995	3082	3211	3330	3481	3762	4104
110	3028	3114	3244	3366	3517	3798	4144
111	3060	3146	3276	3398	3553	3834	4183
112	3092	3179	3312	3434	3586	3870	4223
113	3121	3211	3344	3467	3622	3910	4262
114	3154	3244	3377	3503	3658	3946	4302
115	3186	3276	3409	3535	3690	3982	4342
116	3218	3304	3445	3571	3726	4021	4381
117	3251	3341	3478	3604	3762	4057	4421
118	3283	3373	3510	3640	3798	4093	4460
119	3316	3406	3546	3672	3830	4129	4500
120	3348	3438	3578	3708	3866	4169	4540

(*table continues*)

Table 3-2. Erlang B Traffic Capacity in CCS (*Continued*)

No. of Trunks (*N*)	Traffic (*A*) in CCS for *B* =						
	.001	.002	.005	.010	.020	.050	.100
121	3377	3470	3611	3740	3902	4205	4579
122	3409	3503	3643	3776	3938	4241	4619
123	3442	3535	3679	3809	3971	4280	4658
124	3474	3568	3712	3845	4007	4316	4698
125	3506	3600	3744	3877	4043	4352	4738
126	3539	3632	3780	3913	4079	4392	4777
127	3571	3665	3812	3946	4111	4428	4617
128	3604	3697	3845	3982	4147	4464	4856
129	3636	3733	3881	4014	4183	4504	4896
130	3668	3766	3913	4050	4219	4540	4936
131	3701	3798	3946	4082	4255	4576	4975
132	3733	3830	3978	4118	4288	4615	5015
133	3766	3863	4014	4151	4324	4651	5054
134	3798	3895	4046	4187	4360	4687	5094
135	3830	3928	4079	4219	4396	4727	5134
136	3863	3960	4115	4255	4432	4763	5173
137	3895	3996	4147	4288	4464	4799	5213
138	3928	4028	4183	4324	4500	4835	5252
139	3960	4061	4216	4356	4536	4874	5292
140	3992	4093	4248	4392	4572	4910	5332
141	4025	4126	4284	4424	4608	4946	5371
142	4057	4158	4316	4460	4640	4986	5411
143	4090	4190	4349	4493	4676	5022	5450
144	4122	4226	4385	4529	4712	5058	5490
145	4154	4259	4417	4561	4748	5097	5530
146	4187	4291	4450	4597	4784	5134	5569
147	4219	4324	4486	4630	4817	5170	5609
148	4252	4356	4518	4666	4853	5209	5648
149	4284	4388	4554	4702	4889	5245	5688
150	4316	4424	4586	4738	4925	5281	5728
151	4349	4457	4619	4770	4961	5321	5771
152	4381	4489	4655	4806	4993	5357	5810
153	4414	4522	4687	4842	5029	5393	5850
154	4446	4554	4723	4874	5065	5432	5890
155	4478	4590	4756	4910	5101	5432	5929
156	4511	4622	4788	4946	5137	5504	5969
157	4543	4655	4824	4979	5170	5540	6008
158	4576	4687	4856	5015	5206	5576	6048
159	4608	4720	4892	5051	5242	5616	6088
160	4644	4756	4925	5083	5277	5652	6127

(*table continues*)

Table 3-2. Erlang B Traffic Capacity in CCS (*Continued*)

No. of Trunks (N)	Traffic (A) in CCS for B =						
	.001	.002	.005	.010	.020	.050	.100
161	4676	4788	4957	5119	5314	5688	6167
162	4709	4820	4993	5155	5350	5728	6206
163	4741	4853	5026	5188	5386	5764	6246
164	4774	4889	5062	5224	5418	5800	6286
165	4806	4921	5094	5260	5454	5839	6325
166	4838	4954	5126	5292	5490	5875	6365
167	4871	4986	5162	5328	5526	5911	6404
168	4903	5018	5195	5360	5562	5951	6444
169	4936	5054	5231	5396	5598	5987	6484
170	4972	5087	5263	5429	5634	6026	6523
171	5004	5119	5299	5465	5670	6062	6563
172	5036	5152	5332	5501	5706	6098	6602
173	5069	5188	5368	5533	5738	6138	6642
174	5101	5220	5400	5569	5774	6174	6682
175	5137	5252	5432	5602	5810	6214	6721
176	5170	5288	5468	5638	5846	6250	6761
177	5202	5321	5501	5670	5882	6289	6800
178	5238	5353	5537	5706	5918	6325	6840
179	5267	5386	5569	5738	5954	6365	6880
180	5303	5422	5605	5774	5990	6401	6919
181	5335	5454	5638	5810	6026	6437	6962
182	5368	5486	5674	5843	6062	6476	7002
183	5400	5519	5706	5879	6095	6512	7042
184	5432	5555	5742	5915	6131	6548	7181
185	5465	5587	5774	5947	6167	6588	7121
186	5501	5620	5810	5983	6203	6624	7160
187	5533	5656	5843	6019	6239	6660	7200
188	5566	5688	5879	6052	6275	6700	7240
189	5598	5720	5911	6088	6311	6736	7279
190	5630	5753	5947	6124	6347	6772	7319
191	5666	5785	5980	6156	6383	6811	7358
192	5699	5821	6012	6192	6419	6847	7398
193	5731	5854	6044	6228	6455	6883	7438
194	5767	5890	6080	6260	6491	6923	7477
195	5800	5922	6113	6282	6523	6959	7517
196	5832	5954	6149	6332	6559	6995	7556
197	5864	5987	6181	6365	6595	7034	7596
198	5897	6023	6217	6401	6631	7070	7636
199	5929	6055	6250	6437	6667	7106	7675
200	5962	6091	6286	6469	6703	7146	7715

(table continues)

Table 3-2. Erlang B Traffic Capacity in CCS (*Continued*)

No. of Trunks (N)	Traffic (A) in CCS for B =						
	.001	.002	.005	.010	.020	.050	.100
202	6030	6156	6354	6541	6772	7222	7794
204	6095	6224	6422	6610	6844	7297	7873
206	6163	6293	6494	6678	6916	7369	7956
208	6228	6358	6563	6750	6988	7445	8035
210	6293	6426	6631	6818	7060	7520	8114
212	6361	6494	6700	6890	7132	7596	8194
214	6426	6559	6768	6959	7200	7668	8273
216	6494	6628	6836	7027	7272	7744	8352
218	6559	6696	6905	7099	7344	7819	8431
220	6628	6761	6973	7168	7416	7895	8510
222	6692	6829	7042	7240	7488	7970	8590
224	6761	6898	7110	7308	7560	8042	8672
226	6826	6966	7178	7376	7632	8118	8752
228	6894	7031	7247	7448	7700	8194	8831
230	6959	7099	7315	7517	7772	8269	8910
232	7027	7168	7384	7589	7844	8345	8989
234	7096	7236	7456	7657	7916	8417	9068
236	7160	7301	7524	7729	7988	8492	9148
238	7229	7369	7592	7798	8060	8568	9227
240	7294	7438	7661	7870	8132	8644	9310
242	7362	7506	7729	7938	8204	8719	9389
244	7427	7574	7798	8010	8276	8795	9468
246	7495	7639	7866	8078	8345	8867	9547
248	7564	7708	7934	8145	8420	8942	9626
250	7628	7776	8006	8219	8489	9018	2706
300	9310	9475	9734	9976	10285	10894	11700
350	11005	11189	11473	11743	12085	12773	13694
400	12708	12906	13219	13511	13892	14656	15689
450	14418	14630	14969	15286	15700	16538	17687
500	16135	16362	16722	17064	17510	18425	19681
550	17860	18101	18482	18850	19325	20311	21679
600	19580	19836	20243	20632	21139	22194	23677
650	21312	21586	22010	22421	22957	24084	25675
700	23044	23324	23774	24206	24775	25970	27673
750	24782	25074	25546	25978	26597	27860	29671
800	26518	26694	27313	27784	28415	29750	31669
850	28260	28447	29113	29581	30240	31640	33671
900	29999	30330	30859	31374	32062	33530	35669
950	31745	32087	32638	33170	33887	35413	37670
1000	33491	33844	34412	34963	35708	37296	39672

3.4   HIGH-USAGE TRAFFIC CAPACITY TABLES

High-usage traffic capacity tables, also known as alternate route trunking
tables, are based on the Erlang B distribution. They are used to determine the
traffic carried by each trunk in a trunk group, the cumulative traffic carried by
the trunks, and the overflow traffic to the next trunk or to an alternate route. In
this application, each added trunk carries less traffic than the preceding trunk;
that is, it is less efficient insofar as traffic-carrying capacity is concerned. The
traffic load carried by the last trunk is important in determining the number of
trunks required for a particular high-usage trunk group.

The arrangement of data in high-usage traffic capacity tables implies that
traffic is always offered first to trunk 1, then to trunk 2, and so forth. Modern
switching systems, however, use trunk selection techniques that attempt to
equalize the loading on the individual trunks in the trunk groups. Nevertheless,
the effect on the traffic-carrying capacity of a trunk group by the addition of
another trunk will always be the same.

Tables 3-3 through 3-8 are high-usage traffic capacity tables for trunk
groups containing up to 24 trunks. Telephone operating companies may use
tables applicable to trunk groups containing over a hundred trunks. The
following examples illustrate typical table usage:

Example 3-4

Consider a traffic load of 100 CCS offered to a group of 5 high-usage trunks.
Table 3-3 (pages 50–55) indicates that trunk 1 will carry 26 CCS and overflow
74 CCS to trunk 2. Trunk 2 will carry 23 CCS and overflow 51 CCS to trunk 3.
Trunk 3 will carry 19 CCS and overflow 32 CCS to trunk 4. Trunk 4 will carry
14 CCS and overflow 18 CCS to trunk 5. Referring to Table 3-4 (pages 56–60),
for an offered trunk group traffic load of 100 CCS, trunk 5 will carry 9 CCS and
9 CCS will overflow to an alternate route.

Example 3-5

Determine the traffic load carried by the last trunk, total carried traffic, trunk-
group overflow traffic, and grade of service for a 10-channel trunk group offered
300 CCS of traffic.

In Table 3-5 (pages 61–67), select the *A* row for 300 CCS and read over to the
trunk 10 column to determine that the traffic carried by trunk 10 is 16 CCS, total
carried traffic is 259 CCS, and overflow traffic is 41 CCS.

Grade of service (GOS) = 41 CCS/300 CCS = 0.137.

Example 3-6

For the trunk group of Example 3-5, determine how many trunks must be added to obtain a grade of service of B.1 or better (overflow traffic equals 30 CCS or less).

In Table 3-5, select the *A* row for 300 CCS and read over until 28 CCS overflow traffic is found (28 CCS/300 CCS = 0.093). Read up that column to determine that 11 trunks are required.

Table 3-3. High-Usage Traffic Capacity in CCS (Trunks 1 to 4)

Offered Traffic (A)	Trunk 1 Carried Trunk	Trunk 1 Carried Total	Trunk 1 Over-flow	Trunk 2 Carried Trunk	Trunk 2 Carried Total	Trunk 2 Over-flow	Trunk 3 Carried Trunk	Trunk 3 Carried Total	Trunk 3 Over-flow	Trunk 4 Carried Trunk	Trunk 4 Carried Total	Trunk 4 Over-flow
5	4	4	1									
6	5	5	1									
7	6	6	1									
8	6	6	2									
9	7	7	2									
10	8	8	2									
11	8	8	3									
12	9	9	3									
13	10	10	3									
14	10	10	4									
15	11	11	4	3	14	1						
16	11	11	5	4	15	1						
17	11	11	6	5	16	1						
18	12	12	6	5	17	1						
19	12	12	7	5	17	2						
20	13	13	7	5	18	2						
21	13	13	8	6	19	2						
22	14	14	8	6	20	2						
23	14	14	9	6	20	3						
24	14	14	10	7	21	3						
25	15	15	10	7	22	3						
26	15	15	11	7	22	4						
27	15	15	12	8	23	4						
28	16	16	12	8	24	4						
29	16	16	13	9	25	4						
30	16	16	14	9	25	5	4	29	1			
31	17	17	14	9	26	5	4	30	1			
32	17	17	15	9	26	6	4	30	2			
33	17	17	16	10	27	6	4	31	2			
34	18	18	16	10	28	6	4	32	2			
35	18	18	17	10	28	7	5	33	2			
36	18	18	18	11	29	7	5	34	2			
37	18	18	19	11	29	8	5	34	3			
38	19	19	19	11	30	8	5	35	3			
39	19	19	20	11	30	9	6	36	3			
40	19	19	21	12	31	9	6	37	3			
41	19	19	22	13	32	9	6	38	3			
42	19	19	23	13	32	10	6	38	4			
43	19	19	24	13	32	11	7	39	4			
44	20	20	24	13	33	11	7	40	4			

(*table continues*)

Table 3-3. High-Usage Traffic Capacity in CCS (Trunks 1 to 4 *Continued*)

Offered Traffic (*A*)	Trunk 1 Carried Trunk	Trunk 1 Carried Total	Trunk 1 Over-flow	Trunk 2 Carried Trunk	Trunk 2 Carried Total	Trunk 2 Over-flow	Trunk 3 Carried Trunk	Trunk 3 Carried Total	Trunk 3 Over-flow	Trunk 4 Carried Trunk	Trunk 4 Carried Total	Trunk 4 Over-flow
45	20	20	25	14	34	11	7	41	4			
46	20	20	26	14	34	12	7	41	5			
47	20	20	27	14	34	13	8	42	5			
48	21	21	27	14	35	13	8	43	5			
49	21	21	28	14	35	14	8	43	6			
50	21	21	29	15	36	14	8	44	6			
51	21	21	30	15	36	15	9	45	6	4	49	2
52	21	21	31	15	36	16	9	45	7	5	50	2
53	21	21	32	16	37	16	9	46	7	5	51	2
54	22	22	32	16	37	17	10	47	7	5	52	2
55	22	22	33	16	38	17	10	48	7	5	53	2
56	22	22	34	16	38	18	10	48	8	5	53	3
57	22	22	35	17	39	18	10	49	8	5	54	3
58	22	22	36	17	39	19	10	49	9	6	55	3
59	23	23	36	17	40	19	10	50	9	6	56	3
60	23	23	37	17	40	20	11	51	9	6	57	3
61	23	23	38	17	40	21	11	51	10	6	57	4
62	23	23	39	17	40	22	12	52	10	6	58	4
63	23	23	40	17	40	23	12	52	11	7	59	4
64	23	23	41	18	41	23	12	53	11	7	60	4
65	23	23	42	18	41	24	12	53	12	7	60	5
66	23	23	43	18	41	25	13	54	12	7	61	5
67	23	23	44	19	42	25	13	55	12	7	62	5
68	23	23	45	19	42	26	13	55	13	7	62	6
69	24	24	45	19	43	26	13	56	13	7	63	6
70	24	24	46	19	43	27	13	56	14	8	64	6
71	24	24	47	19	43	28	13	56	15	8	64	7
72	24	24	48	19	43	29	14	57	15	8	65	7
73	24	24	49	20	44	29	14	58	15	8	66	7
74	24	24	50	20	44	30	14	58	16	9	67	7
75	24	24	51	20	44	31	14	58	17	9	67	8
76	24	24	52	20	44	32	15	59	17	9	68	8
77	25	25	52	20	45	32	15	60	17	9	69	8
78	25	25	53	20	45	53	15	60	18	9	69	9
79	25	25	54	20	45	34	15	60	19	10	70	9
80	25	25	55	20	45	35	15	60	20	10	70	10
81	25	25	56	21	46	35	15	61	20	10	71	10
82	25	25	57	21	46	36	16	62	20	10	72	10
83	25	25	58	21	46	37	16	62	21	11	72	11
84	25	25	59	21	46	38	16	62	22	11	73	11

(table continues)

Table 3-3. High-Usage Traffic Capacity in CCS (Trunks 1 to 4 *Continued*)

Offered Traffic (*A*)	Trunk 1 Carried Trunk	Trunk 1 Carried Total	Trunk 1 Over-flow	Trunk 2 Carried Trunk	Trunk 2 Carried Total	Trunk 2 Over-flow	Trunk 3 Carried Trunk	Trunk 3 Carried Total	Trunk 3 Over-flow	Trunk 4 Carried Trunk	Trunk 4 Carried Total	Trunk 4 Over-flow
85	25	25	60	21	46	39	17	63	22	11	74	11
86	25	25	61	21	46	40	17	63	23	11	74	12
87	26	26	61	21	47	40	17	64	23	11	75	12
88	26	26	62	21	47	41	17	64	24	11	75	13
89	26	26	33	22	48	41	17	65	24	11	76	13
90	26	26	64	22	48	42	17	65	25	12	77	13
91	26	26	65	22	48	43	17	65	26	12	77	14
92	26	26	66	22	48	44	17	65	27	12	77	15
93	26	26	67	22	48	45	18	66	27	12	78	15
94	26	26	68	22	48	46	18	66	28	13	79	15
95	26	26	69	23	49	46	18	67	28	13	80	15
96	26	26	70	23	49	47	18	67	29	13	80	16
97	26	26	71	23	49	48	18	67	30	13	80	17
98	26	26	72	23	49	49	19	68	30	13	81	17
99	26	26	73	23	49	50	19	69	31	13	81	18
100	26	26	74	23	49	51	19	68	32	14	82	18
102	27	27	75	23	50	52	19	69	23	14	83	19
104	27	27	77	23	50	54	19	69	35	15	84	20
106	27	27	79	23	50	56	20	70	36	15	85	21
108	27	27	81	24	51	57	20	71	37	15	86	22
110	27	27	83	24	51	59	20	71	39	16	87	23
112	27	27	85	24	51	61	21	72	40	16	88	24
114	27	27	87	24	51	63	21	72	42	16	88	26
116	27	27	89	25	52	64	21	73	43	16	89	27
118	27	27	91	25	52	66	21	73	45	17	90	28
120	28	28	92	25	53	67	21	74	46	17	91	29
122	28	28	94	25	53	69	22	75	47	17	92	30
124	28	28	96	25	53	71	22	75	49	17	92	32
126	28	28	98	25	53	73	22	75	51	18	93	33
128	28	28	100	26	54	74	22	76	52	18	94	34
130	28	28	102	26	54	76	22	76	54	19	95	35
132	28	28	104	26	54	78	23	77	55	19	96	36
134	28	28	106	26	54	80	23	77	57	19	96	38
136	28	28	108	26	54	82	23	77	59	20	97	39
138	29	29	109	26	55	83	23	78	60	20	98	40
140	29	29	111	26	55	85	23	78	62	20	98	42
142	29	29	113	26	55	87	24	79	63	20	99	43
144	29	29	115	26	55	89	24	79	65	20	99	45
146	29	29	117	27	56	90	24	80	66	20	100	46
148	29	29	119	27	56	92	24	80	68	20	100	48

(*table continues*)

Table 3-3. High-Usage Traffic Capacity in CCS (Trunks 1 to 4 *Continued*)

Offered Traffic (A)	Trunk 1 Carried Trunk	Trunk 1 Carried Total	Trunk 1 Over-flow	Trunk 2 Carried Trunk	Trunk 2 Carried Total	Trunk 2 Over-flow	Trunk 3 Carried Trunk	Trunk 3 Carried Total	Trunk 3 Over-flow	Trunk 4 Carried Trunk	Trunk 4 Carried Total	Trunk 4 Over-flow
150	29	29	121	27	56	94	24	80	70	21	101	49
152	29	29	123	27	56	96	25	81	71	21	102	50
154	29	29	125	27	56	98	25	81	73	21	102	52
156	29	29	127	27	56	100	25	81	75	21	102	54
158	30	30	128	27	57	101	25	82	76	21	103	55
160	30	30	130	27	57	103	25	82	78	22	104	56
162	30	30	132	27	57	105	25	82	80	24	104	58
164	30	30	134	27	57	107	25	82	82	22	104	60
166	30	30	136	27	57	109	26	83	83	22	105	61
168	30	30	138	27	57	111	26	83	85	23	106	62
170	30	30	140	27	57	113	26	83	87	23	106	64
172	30	30	142	28	58	114	26	84	88	23	107	65
174	30	30	144	28	58	116	26	84	90	23	107	67
176	30	30	146	28	58	118	26	84	92	24	108	68
178	30	30	148	28	58	120	26	84	94	24	108	70
180	30	30	150	29	59	121	26	85	95	24	109	71
182	30	30	152	29	59	123	26	85	97	24	109	73
184	30	30	154	29	59	125	26	85	99	24	109	75
186	30	30	156	29	59	127	26	85	101	24	109	77
188	30	30	158	29	59	129	27	86	102	24	110	78
190	30	30	160	29	59	131	27	86	104	24	110	80
192	30	30	162	29	59	133	27	86	106	24	110	82
194	30	30	164	29	59	135	27	86	108	25	111	83
196	30	30	166	29	59	137	27	86	110	25	111	85
198	31	31	167	29	60	138	27	87	111	25	112	86
200	31	31	169	29	60	140	27	87	113	25	112	88
202	31	31	171	29	60	142	27	87	115	25	112	90
204	31	31	173	29	60	144	28	88	116	25	113	91
206	31	31	174	29	60	146	28	88	118	25	113	93
208	31	31	177	29	60	148	28	88	120	25	113	95
210	31	31	179	29	60	150	28	88	122	26	114	96
212	31	31	181	29	60	152	28	88	124	26	114	98
214	31	31	183	29	60	154	28	88	126	26	114	100
216	31	31	185	29	60	156	28	88	128	26	114	102
218	31	31	187	30	61	157	28	89	129	26	115	103
220	31	31	189	30	61	159	28	89	131	26	115	105
222	31	31	191	30	61	161	28	89	133	26	115	107
224	31	31	193	30	61	163	28	89	135	27	116	108
226	31	31	195	30	61	165	28	89	137	27	116	110
228	31	31	197	30	61	167	29	90	138	27	117	111

(*table continues*)

Table 3-3. High-Usage Traffic Capacity in CCS (Trunks 1 to 4 *Continued*)

Offered Traffic (*A*)	Trunk 1			Trunk 2			Trunk 3			Trunk 4		
	Carried		Over-flow	Carried		Over-flow	Carried		Over-flow	Carried		Over-flow
	Trunk	Total		Trunk	Total		Trunk	Total		Trunk	Total	
230	31	31	199	30	61	169	29	90	140	27	117	113
232	31	31	201	30	61	171	29	90	142	27	117	115
234	31	31	203	30	61	173	29	90	144	27	117	117
236	31	31	205	30	61	175	29	90	146	27	117	119
238	31	31	207	30	61	177	29	90	148	28	118	120
240	31	31	209	30	61	179	29	90	150	28	118	122
242	31	31	211	30	61	181	29	90	152	28	118	124
244	31	31	213	31	62	182	29	91	153	28	119	125
246	31	31	215	31	62	184	29	91	155	28	119	127
248	31	31	217	31	62	186	29	91	157	28	119	129
250	31	31	219	31	62	188	29	91	159	28	119	131
252	31	31	221	31	62	190	29	91	161	28	119	133
254	31	31	223	31	62	192	29	91	163	28	119	135
256	31	31	225	31	62	194	29	91	165	28	119	137
258	31	31	227	31	62	196	30	92	166	28	120	138
260	31	31	229	31	62	198	30	92	168	28	120	140
262	31	31	231	31	62	200	30	92	170	28	120	142
264	31	31	233	31	62	202	30	92	172	28	120	144
266	32	32	234	31	63	203	30	93	173	28	121	145
268	32	32	236	31	63	205	30	93	175	28	121	147
270	32	32	238	31	63	207	30	93	177	28	121	149
272	32	32	240	31	63	209	30	93	179	28	121	151
274	32	32	242	31	63	211	30	93	181	28	121	153
276	32	32	244	31	63	213	30	93	183	28	121	155
278	32	32	246	31	63	215	30	93	185	29	122	156
280	32	32	248	31	63	217	30	93	187	29	122	158
282	32	32	250	31	63	219	30	93	189	29	122	160
284	32	32	252	31	63	221	30	93	191	29	122	162
286	32	32	254	31	63	223	30	93	193	29	122	164
288	32	32	256	31	63	225	30	93	195	29	122	166
290	32	32	258	31	63	227	31	94	196	29	122	167
292	32	32	260	31	63	229	31	94	198	29	122	169
294	32	32	262	31	63	231	31	94	200	29	122	171
296	32	32	264	31	63	233	31	94	202	29	122	173
298	32	32	266	31	63	235	31	94	204	29	122	175
300	32	32	268	31	63	237	31	94	206	29	122	177
305	32	32	273	32	64	241	31	95	210	29	124	181
310	32	32	278	32	64	246	31	95	215	29	124	186
315	32	32	283	32	64	251	31	95	220	30	125	190
320	32	32	288	32	64	256	31	95	225	30	125	195

(*table continues*)

Table 3-3. High-Usage Traffic Capacity in CCS (Trunks 1 to 4 *Continued*)

Offered Traffic (A)	Trunk 1 Carried Trunk	Trunk 1 Carried Total	Trunk 1 Over-flow	Trunk 2 Carried Trunk	Trunk 2 Carried Total	Trunk 2 Over-flow	Trunk 3 Carried Trunk	Trunk 3 Carried Total	Trunk 3 Over-flow	Trunk 4 Carried Trunk	Trunk 4 Carried Total	Trunk 4 Over-flow
325	32	32	293	32	64	261	31	95	230	30	125	200
330	32	32	298	32	64	266	31	95	235	30	125	205
335	33	33	302	32	65	270	31	96	239	30	126	209
340	33	33	307	32	65	275	31	96	244	30	126	214
345	33	33	312	32	65	280	31	96	249	30	126	219
350	33	33	317	32	65	285	31	96	254	31	127	223
355				32	65	290	31	96	259	31	127	228
360				32	65	295	31	96	264	31	127	233
365				32	65	300	31	96	269	31	127	238
370				32	65	305	32	97	273	31	128	242
375				32	65	310	32	97	278	31	128	247
380				32	65	315	32	97	283	31	128	252
385				32	65	320	32	97	288	31	128	257
390				32	65	325	32	97	293	31	128	262
395				33	66	329	32	98	297	31	129	266
400				33	66	334	32	98	302	31	129	271
405							32	98	307	31	129	276
410							32	98	312	31	129	281
415							32	98	317	32	130	285
420							32	98	322	32	130	290
425							32	98	327	32	130	295
430							32	98	332	32	130	300
435							33	99	336	32	131	304
440							33	99	341	32	131	309
445							33	99	346	32	131	314
450										32	131	319
455										32	131	324
460										32	131	329
465										32	131	334
470										32	131	339

Table 3-4. High-Usage Traffic Capacity in CCS (Trunks 5 to 8)

| Offered Traffic (A) | Trunk 5 | | | Trunk 6 | | | Trunk 7 | | | Trunk 8 | | |
| | Carried | | Over-flow | Carried | | Over-flow | Carried | | Over-flow | Carried | | Over-flow |
	Trunk	Total		Trunk	Total		Trunk	Total		Trunk	Total	
67	3	65	3									
68	4	66	2									
69	4	67	2									
70	4	68	2									
71	4	68	3									
72	4	69	3									
73	4	70	3									
74	4	71	3									
75	5	72	3									
76	5	73	3									
77	5	74	3									
78	5	74	4									
79	5	75	4									
80	6	76	4									
81	6	77	4									
82	6	78	4									
83	6	78	5									
84	6	79	5									
85	6	80	5									
86	7	81	5									
87	7	82	5									
88	7	82	6	4	86	2						
89	7	83	6	4	87	2						
90	7	84	6	4	88	2						
91	7	84	7	4	88	3						
92	8	85	7	4	89	3						
93	8	86	7	4	90	3						
94	8	87	7	4	91	3						
95	8	88	7	4	92	3						
96	8	88	8	5	93	3						
97	8	88	9	5	93	4						
98	8	89	9	5	94	4						
99	8	90	9	5	95	4						
100	9	91	9	5	96	4						
102	9	92	10	6	98	4						
104	9	93	11	6	99	5						
106	10	95	11	6	101	5						
108	10	96	12	6	102	6						
110	10	97	13	7	104	6	3	107	3			
112	11	99	13	7	106	6	3	109	3			

(table continues)

Table 3-4. High-Usage Traffic Capacity in CCS (Trunks 5 to 8 *Continued*)

Offered Traffic (A)	Trunk 5 Carried Trunk	Trunk 5 Carried Total	Trunk 5 Over-flow	Trunk 6 Carried Trunk	Trunk 6 Carried Total	Trunk 6 Over-flow	Trunk 7 Carried Trunk	Trunk 7 Carried Total	Trunk 7 Over-flow	Trunk 8 Carried Trunk	Trunk 8 Carried Total	Trunk 8 Over-flow
114	12	100	14	7	107	7	4	111	3			
116	12	101	15	7	108	8	4	112	4			
118	12	102	16	8	110	8	4	114	4			
120	12	103	17	8	111	9	5	116	4			
122	13	105	17	8	113	9	5	118	4			
124	13	105	19	9	114	10	5	119	5			
126	14	107	19	9	116	10	5	121	5			
128	14	108	20	9	117	11	6	123	5			
130	14	109	21	9	118	12	6	124	6			
132	14	110	22	10	120	12	6	126	6			
134	15	111	23	10	121	13	6	127	7	4	131	3
136	15	112	24	10	122	14	7	129	7	4	133	3
138	15	113	25	11	124	14	7	131	7	4	135	3
140	16	114	26	11	125	15	7	132	8	4	136	4
142	16	115	27	11	126	16	8	134	8	4	138	4
144	16	115	29	12	127	17	8	135	9	5	140	4
146	16	116	30	12	128	18	8	136	10	5	141	5
148	17	117	31	12	129	19	9	138	10	5	143	5
150	17	118	32	13	131	19	9	140	10	5	145	5
152	17	119	33	13	132	20	9	141	11	5	146	6
154	17	119	35	13	133	21	9	142	12	6	148	6
156	18	120	36	14	134	22	9	143	13	6	149	7
158	18	121	37	14	135	23	10	145	13	6	151	7
160	18	122	38	14	136	24	10	146	14	7	153	7
162	19	123	39	14	137	25	10	147	15	7	154	8
164	19	123	41	15	138	26	11	149	15	7	156	8
166	19	124	42	15	139	27	10	150	16	7	157	9
168	19	125	43	15	140	28	11	151	17	8	159	9
170	19	125	45	16	141	29	11	152	18	8	160	10
172	19	126	46	16	142	30	12	154	18	8	162	10
174	20	127	47	16	143	31	12	155	19	8	163	11
176	20	128	48	16	144	32	12	156	20	9	165	11
178	20	128	50	17	145	33	12	157	21	9	166	12
180	20	129	51	17	146	34	12	158	22	9	167	13
182	20	129	53	17	146	36	13	159	23	10	169	13
184	21	130	54	17	147	37	13	160	24	10	170	14
186	21	130	56	18	148	38	14	162	24	10	172	14
188	21	131	57	18	149	39	14	163	25	10	173	15
190	22	132	58	18	150	40	14	164	26	10	174	16
192	22	132	60	18	150	42	15	165	27	10	175	17

(table continues)

Table 3-4. High-Usage Traffic Capacity in CCS (Trunks 5 to 8 *Continued*)

Offered Traffic (*A*)	Trunk 5 Carried Trunk	Trunk 5 Carried Total	Trunk 5 Over-flow	Trunk 6 Carried Trunk	Trunk 6 Carried Total	Trunk 6 Over-flow	Trunk 7 Carried Trunk	Trunk 7 Carried Total	Trunk 7 Over-flow	Trunk 8 Carried Trunk	Trunk 8 Carried Total	Trunk 8 Over-flow
194	22	133	61	18	151	43	15	166	28	11	177	17
196	22	133	63	19	152	44	15	167	29	11	178	18
198	22	134	64	19	153	45	15	168	30	11	179	19
200	22	134	66	19	153	47	16	169	31	12	181	19
202	23	135	67	19	154	48	16	170	32	12	182	20
204	23	136	68	19	155	49	16	171	33	12	183	21
206	23	136	70	19	155	51	16	171	35	13	184	22
208	23	136	72	20	156	52	16	172	36	13	185	23
210	23	137	73	20	157	53	16	173	37	13	186	24
212	23	137	75	20	157	55	17	174	38	13	187	25
214	24	138	76	20	158	56	17	175	39	13	188	26
216	24	138	78	21	159	57	17	176	40	14	190	26
218	24	139	79	21	160	58	17	177	41	14	191	27
220	24	139	81	21	160	60	18	178	42	14	192	28
222	24	139	83	21	160	62	18	178	44	15	193	29
224	24	140	84	21	161	63	18	179	45	15	194	30
226	24	140	86	22	162	64	18	180	46	15	195	31
228	24	141	87	22	163	65	18	181	47	15	196	32
230	24	141	89	22	163	67	19	182	48	15	197	33
232	24	141	91	22	163	69	19	182	50	16	198	34
234	25	142	92	22	164	70	19	183	51	16	199	35
236	25	142	94	22	164	72	20	184	52	16	200	36
238	25	143	95	22	165	73	20	185	53	16	201	37
240	25	143	97	22	165	75	20	185	55	16	201	39
242	25	143	99	23	166	76	20	186	56	17	203	39
244	25	144	100	23	167	77	20	187	57	17	204	40
246	25	144	102	23	167	79	20	187	59	17	204	42
248	25	144	104	24	168	80	20	188	60	17	205	43
250	25	144	106	24	168	82	21	189	61	17	206	44
252	25	144	108	24	168	84	21	189	63	18	207	45
254	26	145	109	24	169	85	21	190	64	18	208	46
256	26	145	111	24	169	87	21	190	66	18	208	48
258	26	146	112	24	170	88	21	191	67	18	209	49
260	26	146	114	24	170	90	22	192	68	18	210	50
262	26	146	116	25	171	91	22	193	69	18	211	51
264	26	146	118	25	171	93	22	193	71	19	212	52
266	26	147	119	25	172	94	22	194	72	19	213	53
268	26	147	121	25	172	96	22	194	74	19	213	55
270	26	147	123	25	172	98	22	194	76	20	214	56
272	27	148	124	25	173	99	22	195	77	20	215	57

(*table continues*)

Table 3-4. High-Usage Traffic Capacity in CCS (Trunks 5 to 8 *Continued*)

Offered Traffic (*A*)	Trunk 5 Carried Trunk	Trunk 5 Carried Total	Trunk 5 Over-flow	Trunk 6 Carried Trunk	Trunk 6 Carried Total	Trunk 6 Over-flow	Trunk 7 Carried Trunk	Trunk 7 Carried Total	Trunk 7 Over-flow	Trunk 8 Carried Trunk	Trunk 8 Carried Total	Trunk 8 Over-flow
274	27	148	126	25	173	101	23	196	78	20	216	58
276	27	148	128	25	173	103	23	196	80	20	216	60
278	27	149	129	25	174	104	23	197	81	20	217	61
280	27	149	131	25	174	106	23	197	83	21	218	62
282	27	149	133	26	175	107	23	198	84	21	219	63
284	27	149	135	26	175	109	23	198	86	21	219	65
286	27	149	137	26	175	111	23	198	88	21	219	67
288	27	149	139	26	175	113	24	199	89	21	220	68
290	27	150	140	26	176	114	24	200	90	21	221	69
292	27	150	142	26	176	116	24	200	92	21	221	71
294	28	151	143	26	177	117	24	201	93	21	222	72
296	28	151	145	26	177	119	24	201	95	21	222	74
298	28	151	147	26	177	121	24	201	97	22	223	75
300	28	151	149	27	178	122	24	202	98	22	224	76
305	28	152	153	27	179	126	24	203	102	22	225	80
310	28	152	158	27	179	131	25	204	106	23	227	83
315	28	153	162	27	180	135	25	205	110	23	228	87
320	29	154	166	27	181	139	25	206	114	23	229	91
325	29	154	171	27	181	144	26	207	118	23	230	95
330	29	154	176	28	182	148	26	208	122	24	232	98
335	29	155	180	28	183	152	26	209	126	24	233	102
340	29	155	185	28	183	157	27	210	130	24	234	106
345	29	155	190	28	183	162	27	210	135	25	235	110
350	29	156	194	28	184	166	27	211	139	25	236	114
355	30	157	198	28	185	170	27	212	143	25	237	118
360	30	157	203	29	186	164	27	213	147	25	238	122
365	30	157	208	29	186	169	27	213	152	26	239	126
370	30	158	212	29	187	183	27	214	156	26	240	130
375	30	158	217	29	187	188	28	215	160	26	241	134
380	30	158	222	29	187	193	28	215	165	27	242	138
385	31	159	226	29	188	197	28	216	169	27	243	142
390	31	159	231	29	188	202	28	216	174	27	243	147
395	31	160	235	29	189	206	28	217	178	27	244	151
400	31	160	240	29	189	211	29	218	182	27	245	155
405	31	160	245	29	189	216	29	218	187	28	246	159
410	31	160	250	30	190	220	29	219	191	28	247	163
415	31	161	254	30	191	224	29	220	195	28	248	167
420	31	161	259	30	191	229	29	220	200	28	248	172
425	31	161	264	30	191	234	29	220	205	28	248	177
430	31	161	269	30	191	239	30	221	209	28	249	181

(table continues)

Table 3-4. High-Usage Traffic Capacity in CCS (Trunks 5 to 8 *Continued*)

Offered Traffic (*A*)	Trunk 5 Carried Trunk	Trunk 5 Carried Total	Trunk 5 Over-flow	Trunk 6 Carried Trunk	Trunk 6 Carried Total	Trunk 6 Over-flow	Trunk 7 Carried Trunk	Trunk 7 Carried Total	Trunk 7 Over-flow	Trunk 8 Carried Trunk	Trunk 8 Carried Total	Trunk 8 Over-flow
435	31	162	273	30	192	243	30	222	213	28	250	185
440	31	162	278	30	192	248	30	222	218	29	251	189
445	31	162	283	30	192	253	30	222	223	29	251	194
450	31	162	288	31	193	257	30	223	227	29	252	198
455	31	162	293	31	193	262	30	223	232	29	252	203
460	32	163	297	31	194	266	30	224	236	29	253	207
465	32	163	302	31	194	271	30	224	241	29	253	212
470	32	163	307	31	194	276	30	224	246	30	254	216
475	32	163	312	31	194	281	30	224	251	30	254	221
480	32	164	316	31	195	285	30	225	255	30	255	225
485	32	165	320	31	196	289	30	226	259	30	256	229
490	32	165	325	31	196	294	30	226	264	30	256	234
495	32	165	330	31	196	299	30	226	269	30	256	239
500	32	165	335	31	196	304	30	226	274	30	256	244
505	32	165	340	31	196	309	31	227	278	30	257	248
510	32	165	345	31	196	314	31	227	283	30	257	253
515	32	165	350	31	196	319	31	227	288	30	257	258
520	32	165	355	32	197	323	31	228	292	30	258	262
525	32	165	360	32	197	328	31	228	297	30	258	267
530				32	197	333	31	228	302	31	259	271
535				32	197	338	31	228	307	31	259	276
540							31	229	311	31	260	280
545							31	229	316	31	260	285
550							31	229	321	31	260	290
555										31	261	294
560										31	261	299
565										31	261	304
570										31	262	308
575										31	262	313
580										31	262	318

Table 3-5. High-Usage Traffic Capacity in CCS (Trunks 9 to 12)

Offered Traffic (A)	Trunk 9 Carried Trunk	Trunk 9 Carried Total	Trunk 9 Over-flow	Trunk 10 Carried Trunk	Trunk 10 Carried Total	Trunk 10 Over-flow	Trunk 11 Carried Trunk	Trunk 11 Carried Total	Trunk 11 Over-flow	Trunk 12 Carried Trunk	Trunk 12 Carried Total	Trunk 12 Over-flow
156	4	153	3									
158	4	155	3									
160	4	157	3									
162	4	158	4									
164	4	160	4									
166	4	161	5									
168	4	163	5									
170	5	165	5									
172	5	167	5									
174	5	168	6									
176	5	170	6									
178	6	172	6									
180	6	173	7									
182	6	175	7	4	179	3						
184	6	176	8	4	180	4						
186	6	178	8	4	182	4						
188	7	180	8	4	184	4						
190	7	181	9	4	185	5						
192	7	182	10	5	187	5						
194	7	184	10	5	189	5						
196	8	186	10	5	191	5						
198	8	187	11	5	192	6						
200	8	189	11	5	194	6						
202	8	190	12	5	195	7						
204	9	192	12	5	197	7						
206	9	193	13	6	199	7	3	202	4			
208	9	194	14	6	200	8	4	204	4			
210	10	196	14	6	202	8	4	206	4			
212	10	197	15	6	203	9	4	207	5			
214	10	198	16	7	205	9	4	209	5			
216	10	200	16	7	207	9	4	211	5			
218	10	201	17	7	208	10	5	213	5			
220	10	202	18	7	209	11	5	214	6			
222	11	204	18	7	211	11	5	216	6			
224	11	205	19	8	213	11	5	218	6			
226	11	206	20	8	214	12	5	219	7			
228	12	208	20	8	216	12	5	221	7			
230	12	209	21	8	217	13	6	223	7			
232	12	210	22	9	219	13	6	225	7	3	228	4
234	12	211	23	9	220	14	6	226	8	4	230	4

(*table continues*)

Table 3-5. High-Usage Traffic Capacity in CCS (Trunks 9 to 12 *Continued*)

Offered Traffic (A)	Trunk 9 Carried Trunk	Trunk 9 Carried Total	Trunk 9 Over-flow	Trunk 10 Carried Trunk	Trunk 10 Carried Total	Trunk 10 Over-flow	Trunk 11 Carried Trunk	Trunk 11 Carried Total	Trunk 11 Over-flow	Trunk 12 Carried Trunk	Trunk 12 Carried Total	Trunk 12 Over-flow
236	12	212	24	9	221	15	6	227	9	4	231	5
238	13	214	24	9	223	15	6	229	9	4	233	5
240	13	214	26	10	224	16	7	231	9	4	235	5
242	13	216	26	10	226	16	7	233	9	4	237	5
244	13	217	27	10	227	17	7	234	10	4	238	6
246	14	218	28	10	228	18	7	235	11	5	240	6
248	14	219	29	11	230	18	7	237	11	5	242	6
250	14	220	30	11	231	19	7	238	12	5	243	7
252	14	221	31	11	232	20	8	240	12	5	245	7
254	14	222	32	11	233	21	8	241	13	6	247	7
256	15	223	33	12	235	21	8	243	13	6	249	7
258	15	224	34	12	236	22	8	244	14	6	250	8
260	15	225	35	12	237	23	9	246	14	6	252	8
262	15	226	36	12	238	24	9	247	15	6	253	9
264	15	227	37	12	239	25	9	248	16	7	255	9
266	16	229	37	12	241	25	9	250	16	7	257	9
268	16	229	39	13	242	26	9	251	17	7	258	10
270	16	230	40	13	243	27	10	253	17	7	260	10
272	16	231	41	13	244	28	10	254	18	7	261	11
274	16	232	42	14	246	28	10	256	18	7	263	11
276	17	233	43	14	247	29	10	257	19	7	264	12
278	17	234	44	14	248	30	10	258	20	8	266	12
280	17	235	45	14	249	31	10	259	21	8	267	13
282	17	236	46	14	250	32	11	261	21	8	269	13
284	17	236	48	15	251	33	11	262	22	8	270	14
286	18	237	49	15	252	34	11	263	23	8	271	15
288	18	238	50	15	253	35	11	264	24	9	273	15
290	18	239	51	15	254	36	12	266	24	9	275	15
292	19	238	52	15	255	37	12	267	25	9	276	16
294	19	241	53	15	256	38	12	268	26	10	278	16
296	19	241	55	16	257	39	12	269	27	10	279	17
298	19	242	56	16	258	40	12	270	28	10	280	18
300	19	243	57	16	259	41	13	272	28	10	282	18
305	20	245	60	16	261	44	14	275	30	10	285	20
310	20	247	63	17	264	46	14	278	32	10	288	22
315	20	248	67	18	266	49	14	280	35	11	291	24
320	21	250	70	18	268	52	15	283	37	11	294	26
325	21	251	74	19	270	55	16	286	39	12	298	27
330	21	253	77	19	272	58	16	288	42	13	301	29
335	22	255	80	19	274	61	16	290	45	13	303	32

(*table continues*)

Table 3-5. High-Usage Traffic Capacity in CCS (Trunks 9 to 12 *Continued*)

Offered Traffic (*A*)	Trunk 9 Carried Trunk	Trunk 9 Carried Total	Trunk 9 Over-flow	Trunk 10 Carried Trunk	Trunk 10 Carried Total	Trunk 10 Over-flow	Trunk 11 Carried Trunk	Trunk 11 Carried Total	Trunk 11 Over-flow	Trunk 12 Carried Trunk	Trunk 12 Carried Total	Trunk 12 Over-flow
340	22	256	84	19	275	65	17	292	48	14	306	34
345	23	258	87	20	278	67	17	295	50	14	309	36
350	23	259	91	20	279	71	18	297	53	14	311	39
355	23	260	95	21	281	74	18	299	56	15	314	41
360	24	262	98	21	283	77	18	301	59	16	317	43
365	24	263	102	21	284	81	19	303	62	16	319	46
370	24	264	106	22	286	84	19	305	65	17	322	48
375	24	265	110	22	287	88	20	307	68	17	324	51
380	25	267	113	22	289	91	20	309	71	17	326	54
385	25	268	117	22	290	95	20	310	75	18	328	57
390	26	269	121	23	292	98	20	312	78	18	330	60
395	26	270	125	23	293	102	21	314	81	19	333	62
400	26	271	129	23	294	106	22	316	84	19	335	65
405	26	272	133	23	295	110	22	317	88	20	337	68
410	26	273	137	24	297	113	22	319	91	20	339	71
415	26	274	141	24	298	117	22	320	95	20	340	75
420	26	274	146	25	299	121	23	322	98	20	342	78
425	27	275	150	25	300	125	23	323	102	21	344	81
430	27	276	154	25	301	129	24	325	105	21	346	84
435	27	277	158	25	302	133	24	326	109	21	347	88
440	27	278	162	25	303	137	24	327	113	22	349	91
445	27	278	167	26	304	141	24	328	117	22	350	95
450	27	279	171	26	305	145	24	329	121	23	352	98
455	27	280	175	26	306	149	24	330	125	23	353	102
460	28	281	179	26	307	153	25	332	128	23	355	105
465	28	281	184	27	308	157	25	333	132	23	356	109
470	28	282	188	27	309	161	25	334	136	23	357	113
475	28	282	193	27	309	166	26	335	140	23	358	117
480	28	283	197	27	310	170	26	336	144	24	360	120
485	28	284	201	27	311	174	26	337	148	24	361	124
490	28	284	206	28	312	178	26	338	152	24	362	128
495	29	285	210	28	313	182	26	339	156	24	363	132
500	29	285	215	28	313	187	27	340	160	25	365	135
505	29	286	219	28	314	191	27	341	164	25	366	139
510	30	287	223	28	315	195	27	342	168	25	367	143
515	30	287	228	28	315	200	27	342	173	26	368	147
520	30	288	232	28	316	204	27	343	177	26	369	151
525	30	288	237	28	316	209	28	344	181	26	370	155
530	30	289	241	28	317	213	28	345	185	26	371	159
535	30	289	246	28	317	218	28	345	190	27	372	163

(*table continues*)

Table 3-5. High-Usage Traffic Capacity in CCS (Trunks 9 to 12 *Continued*)

Offered Traffic (*A*)	Trunk 9 Carried Trunk	Trunk 9 Carried Total	Trunk 9 Over-flow	Trunk 10 Carried Trunk	Trunk 10 Carried Total	Trunk 10 Over-flow	Trunk 11 Carried Trunk	Trunk 11 Carried Total	Trunk 11 Over-flow	Trunk 12 Carried Trunk	Trunk 12 Carried Total	Trunk 12 Over-flow
540	30	290	250	28	318	222	28	346	194	27	373	167
545	30	290	255	29	319	226	28	347	198	27	374	171
550	30	290	260	29	319	231	28	347	203	27	374	176
555	30	291	264	29	320	235	28	348	207	27	375	180
560	30	291	269	29	320	240	28	348	212	28	376	184
565	30	291	274	30	321	244	28	349	216	28	377	188
570	30	292	278	30	322	248	28	350	220	28	378	192
575	30	292	283	30	322	253	29	351	224	28	379	196
580	30	292	288	30	322	258	29	351	229	28	379	201
585	30	293	292	30	323	262	29	352	233	28	380	205
590	31	294	296	30	324	266	29	353	237	28	381	209
595	31	294	301	30	324	271	30	354	241	28	382	213
600	31	294	306	30	324	276	30	354	246	28	382	218
605	31	294	311	30	324	281	30	354	251	29	383	222
610	31	295	315	30	325	285	30	355	255	29	384	226
615	31	296	319	30	326	289	30	356	259	29	385	230
620	31	296	324	30	326	294	30	356	264	29	385	235
625	31	296	329	30	326	299	30	356	269	29	385	240
630	31	296	334	31	327	303	30	357	273	29	386	244
635	31	296	339	31	327	308	30	357	278	29	386	249
640	31	297	343	31	328	312	30	358	282	29	387	253
645	31	297	348	31	328	317	30	358	287	30	388	257
650	31	297	353	31	328	322	30	358	292	30	388	262
655	31	298	357	31	329	326	30	359	296	30	389	266
660	31	298	362	31	329	331	30	359	301	30	389	271
665				31	330	335	30	360	305	30	390	275
670				31	330	340	30	360	310	30	390	280
675				31	330	345	31	361	314	30	391	284
680				31	330	350	31	361	319	30	391	289
685				31	330	355	31	361	324	30	391	294
690							31	362	328	30	392	298
695							31	362	333	30	392	303
700							31	363	337	30	393	307
705							31	363	342	30	393	312
710										30	394	316

Table 3-6. High-Usage Traffic Capacity in CCS (Trunks 13 to 16)

Offered Traffic (A)	Trunk 13 Carried Trunk	Trunk 13 Carried Total	Trunk 13 Over-flow	Trunk 14 Carried Trunk	Trunk 14 Carried Total	Trunk 14 Over-flow	Trunk 15 Carried Trunk	Trunk 15 Carried Total	Trunk 15 Over-flow	Trunk 16 Carried Trunk	Trunk 16 Carried Total	Trunk 16 Over-flow
258	4	254	4									
260	4	256	4									
262	4	257	5									
264	4	259	5									
266	4	261	5									
268	4	262	6									
270	4	264	6									
272	5	266	6									
274	5	268	6									
276	5	269	7									
278	5	271	7									
280	5	272	8									
282	5	274	8	4	278	4						
284	6	276	8	4	280	4						
286	6	277	9	4	281	5						
288	6	279	9	4	283	5						
290	6	281	9	4	285	5						
292	6	282	10	4	286	6						
294	6	284	10	4	288	6						
296	6	285	11	5	290	6						
298	7	287	11	5	292	6						
300	7	289	11	5	294	6						
305	7	292	13	5	297	8						
310	8	296	14	6	302	8	4	306	4			
315	9	300	15	6	306	9	4	310	5			
320	9	303	17	7	310	10	4	314	6			
325	9	307	18	7	314	11	4	318	7			
330	10	311	19	7	318	12	5	323	7			
335	11	314	21	8	322	13	5	327	8			
340	11	317	23	8	325	15	6	331	9	4	335	5
345	11	320	25	9	329	16	6	335	10	4	339	6
350	12	323	27	9	332	18	7	339	11	5	344	6
355	12	326	29	10	336	19	7	343	12	5	348	7
360	13	330	30	10	340	20	7	347	13	5	352	8
365	13	332	33	11	343	22	7	350	15	6	356	9
370	13	335	35	11	346	24	8	354	16	6	360	10
375	14	338	37	11	349	26	9	358	17	6	364	11
380	15	341	39	12	353	27	9	362	18	6	368	12
385	15	343	42	13	356	29	9	365	20	7	372	13
390	16	346	44	13	359	31	9	368	22	8	376	14

(table continues)

Table 3-6. High-Usage Traffic Capacity in CCS (Trunks 13 to 16 *Continued*)

Offered Traffic (*A*)	Trunk 13 Carried Trunk	Trunk 13 Carried Total	Trunk 13 Over-flow	Trunk 14 Carried Trunk	Trunk 14 Carried Total	Trunk 14 Over-flow	Trunk 15 Carried Trunk	Trunk 15 Carried Total	Trunk 15 Over-flow	Trunk 16 Carried Trunk	Trunk 16 Carried Total	Trunk 16 Over-flow
395	16	349	46	13	362	33	10	372	23	8	380	15
400	16	351	49	13	364	36	11	375	25	9	384	16
405	16	353	52	14	367	38	11	378	27	9	387	18
410	17	356	54	14	370	40	12	382	28	9	391	19
415	18	358	57	15	373	42	12	385	30	9	394	21
420	18	360	60	15	375	45	13	388	32	10	398	22
425	18	362	63	16	378	47	13	391	34	10	401	24
430	18	364	66	16	380	50	14	394	36	11	405	25
435	19	366	69	17	383	52	14	397	38	11	408	27
440	19	368	72	17	385	55	14	399	41	12	411	29
445	20	370	75	17	387	58	15	402	43	12	414	31
450	20	372	78	18	390	60	15	405	45	12	417	33
455	21	374	81	18	392	63	15	407	48	13	420	35
460	21	376	84	18	394	66	16	410	50	13	423	37
465	21	377	88	19	396	69	16	412	53	14	426	39
470	22	379	91	19	398	72	17	415	55	14	429	41
475	22	380	95	20	400	75	17	417	58	14	431	44
480	22	382	98	20	402	78	17	419	61	15	434	46
485	23	384	101	20	404	81	18	422	63	15	437	48
490	23	385	105	21	406	84	18	424	66	16	440	50
495	23	386	109	21	407	88	19	426	69	16	442	53
500	23	388	112	21	409	91	19	428	72	16	444	56
505	23	389	106	22	411	94	19	430	75	17	447	58
510	23	390	120	22	412	98	20	432	78	17	449	61
515	24	392	123	22	414	101	20	434	81	17	451	64
520	24	393	127	23	416	104	20	436	84	18	454	66
525	24	394	131	23	417	108	21	438	87	18	456	69
530	24	395	135	23	418	112	21	439	91	19	458	72
535	25	397	138	23	420	115	21	441	94	19	460	75
540	25	398	142	23	421	119	22	443	97	19	462	78
545	25	399	146	23	422	123	22	444	101	20	464	81
550	26	400	150	24	424	126	22	446	104	20	466	84
555	26	401	154	24	425	130	22	447	108	21	468	87
560	26	402	158	24	426	134	22	448	112	21	469	91
565	26	403	162	24	427	138	23	450	115	21	471	94
570	26	404	166	25	429	141	23	452	118	21	473	97
575	26	405	170	25	430	145	23	453	122	22	475	100
580	26	406	174	25	431	149	23	454	126	22	476	104
585	26	407	178	25	432	153	24	456	129	22	478	107
590	27	408	182	25	433	157	24	457	133	23	480	110

(*table continues*)

Table 3-6. High-Usage Traffic Capacity in CCS (Trunks 13 to 16 *Continued*)

Offered Traffic (*A*)	Trunk 13 Carried Trunk	Trunk 13 Carried Total	Trunk 13 Over-flow	Trunk 14 Carried Trunk	Trunk 14 Carried Total	Trunk 14 Over-flow	Trunk 15 Carried Trunk	Trunk 15 Carried Total	Trunk 15 Over-flow	Trunk 16 Carried Trunk	Trunk 16 Carried Total	Trunk 16 Over-flow
595	27	409	186	25	434	161	24	458	137	23	481	114
600	27	409	191	26	435	165	24	459	141	23	482	118
605	27	410	195	26	436	169	25	461	144	23	484	121
610	27	411	199	26	437	173	25	462	148	23	485	125
615	27	412	203	26	438	177	25	463	152	23	486	129
620	28	413	207	26	439	181	26	465	155	23	488	132
625	28	413	212	27	440	185	26	466	159	24	490	135
630	28	414	216	27	441	189	26	467	163	24	491	139
635	29	415	220	27	442	193	26	468	167	24	492	143
640	29	416	224	27	443	197	26	469	171	24	493	147
645	29	417	228	27	444	201	26	470	175	25	495	150
650	29	417	233	27	444	206	27	471	179	25	496	154
655	29	418	237	27	445	210	27	472	183	25	497	158
660	29	418	242	28	446	214	27	473	187	25	498	162
665	29	419	246	28	447	218	27	474	191	25	499	166
670	29	419	251	28	447	223	27	474	196	26	500	170
675	29	420	255	28	448	227	27	475	200	26	501	174
680	29	421	259	28	449	231	27	476	204	26	502	178
685	29	421	264	28	449	236	28	477	208	26	503	182
690	30	422	268	28	450	240	28	478	212	26	504	186
695	30	422	273	29	451	244	28	479	216	26	505	190
700	30	423	277	29	452	248	28	480	220	26	506	194
705	30	423	282	29	452	253	28	480	225	27	507	198
710	30	424	286	29	453	257	28	481	229	27	508	202
715	30	425	290	29	454	261	28	482	233	27	509	206
720	30	425	295	29	454	266	28	482	238	27	509	211
725	30	425	300	29	454	271	29	483	242	27	510	215
730	30	426	304	29	455	275	29	484	246	27	511	219
735	30	426	309	29	455	280	29	484	251	28	512	223
740	30	427	313	29	456	284	29	485	255	28	513	227
745	30	427	318	30	457	288	29	486	259	28	514	231
750	30	428	322	30	458	292	29	487	263	28	515	235
755	30	428	327	30	458	297	29	487	268	28	515	240
760	30	428	332	30	458	302	29	487	273	29	516	244
765				30	459	306	29	488	277	29	517	248
770				30	459	311	29	488	282	29	517	253
775				30	460	315	29	489	286	29	518	257
780				30	460	320	30	490	290	29	519	261
785				30	460	325	30	490	295	29	519	266
790							30	491	299	29	520	270

Table 3-7. High-Usage Traffic Capacity in CCS (Trunks 17 to 20)

Offered Traffic (A)	Trunk 17 Carried Trunk	Trunk 17 Carried Total	Trunk 17 Over-flow	Trunk 18 Carried Trunk	Trunk 18 Carried Total	Trunk 18 Over-flow	Trunk 19 Carried Trunk	Trunk 19 Carried Total	Trunk 19 Over-flow	Trunk 20 Carried Trunk	Trunk 20 Carried Total	Trunk 20 Over-flow
365	4	360	5									
370	4	364	6									
375	4	368	7									
380	5	373	7									
385	5	377	8									
390	5	381	9	4	385	5						
395	5	385	10	4	389	6						
400	6	390	10	4	394	6						
405	6	393	12	5	398	7						
410	7	397	13	5	402	8						
415	7	401	14	5	406	9						
420	7	405	15	6	411	9	4	415	5			
425	8	409	16	6	415	10	4	419	6			
430	8	413	17	6	419	11	4	423	7			
435	9	417	18	6	423	12	5	428	7			
440	9	420	20	7	427	13	5	432	8			
445	10	424	21	7	431	14	5	436	9	4	440	5
450	10	427	23	8	435	15	5	440	10	4	444	6
455	10	430	25	8	438	17	6	444	11	4	448	7
460	11	434	26	8	442	18	6	448	12	4	452	8
465	11	437	28	9	446	19	6	452	13	5	457	8
470	11	440	30	9	449	21	7	456	14	5	461	9
475	12	443	32	10	453	22	7	460	15	5	465	10
480	12	446	34	10	456	24	8	464	16	5	469	11
485	13	450	35	10	460	25	8	468	17	6	474	11
490	13	453	37	11	464	26	8	472	18	6	478	12
495	13	455	40	11	466	29	9	475	20	7	482	13
500	14	458	42	12	470	30	9	479	21	7	486	14
505	14	461	44	12	473	32	10	483	22	7	490	15
510	15	464	46	12	476	34	10	486	24	7	493	17
515	15	466	49	13	479	36	10	489	26	8	497	18
520	15	469	51	13	482	38	11	493	27	8	501	19
525	16	472	53	13	485	40	11	496	29	9	505	20
530	16	474	56	14	488	42	11	499	31	9	508	22
535	17	477	58	14	491	44	12	503	32	9	512	23
540	17	479	61	15	494	46	12	506	34	10	516	24
545	17	481	64	15	496	49	13	509	36	10	519	26
550	18	484	66	15	499	51	13	512	38	10	522	28
555	18	486	69	16	502	53	13	515	40	11	526	29
560	19	488	72	16	504	56	14	518	42	11	529	31

(table continues)

Table 3-7. High-Usage Traffic Capacity in CCS (Trunks 17 to 20 *Continued*)

Offered Traffic (*A*)	Trunk 17			Trunk 18			Trunk 19			Trunk 20		
	Carried		Over-flow	Carried		Over-flow	Carried		Over-flow	Carried		Over-flow
	Trunk	Total		Trunk	Total		Trunk	Total		Trunk	Total	
565	19	490	75	16	506	59	14	520	45	12	532	33
570	19	492	78	17	509	61	14	523	47	12	535	35
575	19	494	81	17	511	64	15	526	49	12	538	37
580	20	496	84	17	513	67	15	528	52	13	541	39
585	20	498	87	18	516	69	15	531	54	13	544	41
590	20	500	90	18	518	72	16	534	56	13	547	43
595	21	502	93	18	520	75	16	536	59	14	550	45
600	21	503	97	19	522	78	17	539	61	14	553	47
605	21	505	100	19	524	81	17	541	64	15	556	49
610	22	507	103	19	526	84	17	543	67	15	558	52
615	22	508	107	20	528	87	18	546	69	15	561	54
620	22	510	110	20	530	90	18	548	72	16	564	56
625	22	512	113	20	532	93	18	550	75	16	566	59
630	22	513	117	21	534	96	18	552	78	17	569	61
635	23	515	120	21	536	99	18	554	81	17	571	64
640	23	516	124	21	537	103	19	556	84	17	573	67
645	23	518	127	21	539	106	19	558	87	18	576	69
650	23	519	131	21	540	110	20	560	90	18	578	72
655	23	520	135	22	542	113	20	562	93	18	580	75
660	24	522	138	22	544	116	20	564	96	19	583	77
665	24	523	142	23	546	119	20	566	99	19	585	80
670	24	524	146	23	547	123	21	568	102	19	587	83
675	25	526	149	23	549	126	21	570	105	19	589	86
680	25	527	153	23	550	130	21	571	109	20	591	89
685	25	528	157	24	552	133	21	573	112	20	593	92
690	25	529	161	24	553	137	22	575	115	20	595	95
695	26	531	164	24	555	140	22	577	118	20	597	98
700	26	532	168	24	556	144	22	578	122	20	598	102
705	26	533	172	24	557	148	22	579	126	21	600	105
710	26	534	176	24	558	152	23	581	129	21	602	108
715	26	535	180	25	560	155	23	583	132	21	604	111
720	27	536	184	25	561	159	23	584	136	21	605	115
725	27	537	188	25	562	163	23	585	140	22	607	118
730	27	538	192	25	563	167	24	587	143	22	609	121
735	27	539	196	25	564	171	24	588	147	22	610	125
740	27	540	200	25	565	175	24	589	151	23	612	128
745	27	541	204	26	567	178	24	591	154	23	614	131
750	27	542	208	26	568	182	24	592	158	23	615	135
755	27	543	212	26	569	186	24	593	162	23	616	139
760	27	544	216	26	570	190	24	594	166	24	618	142

(*table continues*)

Table 3-7. High-Usage Traffic Capacity in CCS (Trunks 17 to 20 *Continued*)

Offered Traffic (A)	Trunk 17 Carried		Over-flow	Trunk 18 Carried		Over-flow	Trunk 19 Carried		Over-flow	Trunk 20 Carried		Over-flow
	Trunk	Total		Trunk	Total		Trunk	Total		Trunk	Total	
765	28	545	220	26	571	194	25	596	169	24	620	145
770	28	545	225	27	572	198	25	597	173	24	621	149
775	28	546	229	27	573	202	25	598	177	24	622	153
780	28	547	233	27	574	206	25	599	181	24	623	157
785	28	547	238	27	574	211	26	600	185	25	625	160
790	28	548	242	27	575	215	26	601	189	25	626	164
795	28	549	246	27	576	219	26	602	193	25	627	168
800	29	550	250	27	577	223	26	603	197	25	628	172
810	29	551	259	28	579	231	26	605	205	25	631	179
820	29	552	268	28	580	240	27	607	213	26	633	187
830	29	554	276	28	582	248	27	609	221	26	635	195
840	29	555	285	28	583	257	28	611	229	26	637	203
850	29	556	294	29	585	265	28	613	237	26	632	211
860	29	557	303	29	586	274	28	614	246	27	641	219
870	30	559	311	29	588	282	28	616	254	27	643	227
880	30	560	320	29	589	291	28	617	263	28	645	235
890	30	561	329	29	590	300	28	618	272	28	646	244
900	30	562	338	29	591	309	29	620	280	28	648	252
910				30	593	317	29	622	288	28	650	260
920				30	594	326	29	623	297	28	651	269

Table 3-8. High-Usage Traffic Capacity in CCS (Trunks 21 to 24)

Offered Traffic (A)	Trunk 21 Carried Trunk	Trunk 21 Carried Total	Trunk 21 Over-flow	Trunk 22 Carried Trunk	Trunk 22 Carried Total	Trunk 22 Over-flow	Trunk 23 Carried Trunk	Trunk 23 Carried Total	Trunk 23 Over-flow	Trunk 24 Carried Trunk	Trunk 24 Carried Total	Trunk 24 Over-flow
470	4	465	5									
475	4	469	6									
480	4	473	7									
485	4	478	7									
490	4	482	8									
495	4	486	9									
500	5	491	9	4	495	5						
505	5	495	10	4	499	6						
510	6	499	11	4	503	7						
515	6	503	12	4	507	8						
520	6	507	13	5	512	8						
525	6	511	14	5	516	9						
530	7	515	15	5	520	10	4	524	6			
535	7	519	16	5	524	11	4	528	7			
540	7	523	17	6	529	11	4	533	7			
545	8	527	18	6	533	12	4	537	8			
550	8	530	20	7	537	13	4	541	9			
555	8	534	21	7	541	14	5	546	9			
560	9	538	22	7	545	15	5	550	10	4	554	6
565	9	541	24	7	548	17	6	554	11	4	558	7
570	10	545	25	7	552	18	6	558	12	4	562	8
575	10	548	27	8	556	19	6	562	13	5	567	8
580	11	552	28	8	560	20	6	566	14	5	571	9
585	11	555	30	9	564	21	6	570	15	5	575	10
590	11	558	32	9	567	23	7	574	16	5	579	11
595	12	562	33	9	571	24	7	578	17	6	584	11
600	12	565	35	10	575	25	7	582	18	6	588	12
605	12	568	37	10	578	27	8	586	19	6	592	13
610	13	571	39	11	582	28	8	590	20	6	596	14
615	13	574	41	11	585	30	8	593	22	7	600	15
620	13	577	43	11	588	32	9	597	23	7	604	16
625	14	580	45	12	592	33	9	601	24	7	608	17
630	14	583	47	12	595	35	9	604	26	8	612	18
635	15	586	49	12	598	37	10	608	27	8	616	19
640	15	588	52	13	601	39	10	611	29	8	619	21
645	15	591	54	13	604	41	10	614	31	9	623	22
650	16	594	56	13	607	43	11	618	32	9	627	23
655	16	596	59	14	610	45	11	621	34	9	630	25
660	16	599	61	14	613	47	11	624	36	10	634	26
665	16	601	64	14	615	47	12	627	36	10	637	28

(*table continues*)

Table 3-8. High-Usage Traffic Capacity in CCS (Trunks 21 to 24 *Continued*)

Offered Traffic (A)	Trunk 21 Carried Trunk	Trunk 21 Carried Total	Trunk 21 Over-flow	Trunk 22 Carried Trunk	Trunk 22 Carried Total	Trunk 22 Over-flow	Trunk 23 Carried Trunk	Trunk 23 Carried Total	Trunk 23 Over-flow	Trunk 24 Carried Trunk	Trunk 24 Carried Total	Trunk 24 Over-flow
670	17	604	66	14	618	52	13	631	39	10	641	29
675	17	606	69	15	621	54	13	634	41	10	644	31
680	17	608	72	15	623	57	13	636	44	11	647	33
685	18	611	74	16	627	58	13	640	45	11	651	34
690	18	613	77	16	629	61	13	642	48	12	654	36
695	18	615	80	16	631	64	14	645	50	12	657	38
700	19	617	83	17	634	66	14	648	52	12	660	40
705	19	619	86	17	636	69	15	651	54	12	663	42
710	20	622	88	17	639	71	15	654	56	13	667	43
715	20	624	91	17	641	74	15	656	59	13	669	46
720	20	625	95	18	643	77	16	659	61	14	673	47
725	20	627	98	18	645	80	16	661	64	14	675	50
730	20	629	101	19	648	82	16	664	66	14	678	52
735	21	631	104	19	650	85	17	667	68	14	681	54
740	21	633	107	19	652	88	17	669	71	15	684	56
745	21	635	110	19	654	91	17	671	74	15	686	59
750	21	636	114	20	656	94	18	674	76	15	689	61
755	21	638	117	20	658	97	18	676	79	16	692	63
760	21	640	120	20	660	100	18	678	82	16	694	66
765	22	642	123	20	662	103	18	680	85	17	697	68
770	22	643	127	20	663	107	19	682	88	17	699	71
775	22	644	131	21	665	110	19	684	91	17	701	74
780	23	646	134	21	667	113	19	686	94	18	704	76
785	23	648	137	21	669	116	19	688	97	18	706	79
790	23	649	141	21	670	120	20	690	100	18	708	82
795	24	651	144	21	672	123	20	692	103	19	711	84
800	24	652	148	22	674	126	20	694	106	19	713	87
810	24	655	155	22	677	133	21	698	112	19	717	93
820	24	657	163	23	680	140	22	702	118	19	721	99
830	25	660	170	23	683	147	22	705	125	20	725	105
840	25	662	178	24	686	154	22	708	132	21	729	111
850	26	665	185	24	689	161	23	712	138	21	733	117
860	26	667	193	25	692	168	23	715	145	21	736	124
870	26	669	201	25	694	176	24	718	152	22	740	130
880	26	671	209	25	696	184	24	720	160	23	743	137
890	27	673	217	26	699	191	24	723	167	23	746	144
900	27	675	225	26	701	199	25	726	174	23	749	151
910	27	677	233	26	703	207	25	728	182	24	752	158
920	28	679	241	26	705	215	25	730	190	24	754	166
930	28	681	249	26	707	223	26	733	197	24	757	173

3.5   PEAKED TRAFFIC CAPACITY TABLES

Peaked traffic capacity tables, also known as Wilkinson or Neal-Wilkinson high-usage trunk capacity tables, are based on the Erlang B distribution and the Neal-Wilkinson equivalent random theory.   They are used in the North American PSTN in lieu of the Poisson traffic capacity tables to compensate for system-induced peakedness caused by alternate-routing overflow traffic. Peakedness in offered traffic can also be caused by abnormally high traffic, such as on Mother's Day and Christmas, or due to a disaster, stock market crash, or similar occurrence.

The peakedness factor (Z) for any trunk group is obtained by calculating the variance-to-mean ratio of the busy-hour traffic, as expressed in Equation 3-2.   Peakedness factors of unity (one) indicate random (Poisson) traffic, whereas factors greater than unity indicate peaked traffic.   Traffic with a peakedness factor less than unity indicates smooth traffic, which is treated as random traffic.

(3.2)

$$Z = \frac{v}{\alpha}$$

where Z = peakedness factor

v = variance of the traffic data $= \sum \frac{(x - \alpha)^2}{(N - 1)}$

α = arithmetic mean (average) of the sample values $= \frac{\sum x}{N}$

x = individual values in the sample

N = number of values in the sample

Tables 3-9 (page 75) and 3.10 (page 76) are used for trunk groups engineered for a grade of service of B.01. Tables 3-11 (page 77) and 3-12 (page 78) are used for trunk groups engineered for a grade of service of B.02.   The following example illustrates typical peakedness-factor determination and peaked traffic capacity table usage.

Example 3-6

Determine the peakedness factor and the number of trunks required for a final trunk group in an alternate routing system engineered for B.02 service, where the busy-hour traffic statistics are as follows:

Sample	x (CCS)	$x - \alpha$	$(x - \alpha)^2$
1	115	-14.4	207.4
2	103	-26.4	697.0
3	145	+15.6	243.4
4	140	+10.6	112.4
5	131	+1.6	2.6
6	147	+17.6	309.8
7	142	+12.6	158.8
8	112	-17.4	302.8
	$\Sigma x = 1035$		$\Sigma(x - \alpha)^2 = 2034.2$

$$\alpha = \frac{\Sigma x}{N} = \frac{1035}{8} = 129.4 \text{ CCS}$$

$$v = \Sigma \frac{(x - \alpha)^2}{(N - 1)} = \frac{(2034.2)}{7} = 290.6$$

$$Z = \frac{v}{\alpha} = \frac{290.6}{129.4} = 2.2$$

Referring to Table 3-11, determine that 12 trunks are required to handle up to 139 CCS of offered traffic for a peakedness factor of 2.2.

Table 3-9. B.01 Peaked Traffic Capacity in CCS (Z = 1.0 to 2.4)

No. of Trunks (N)	Traffic (A) in CCS for Z =							
	1.0	1.2	1.4	1.6	1.8	2.0	2.2	2.4
2	5							
4	31	19	8					
6	69	55	41	29	17	4		
8	113	97	81	69	55	43	29	15
10	161	142	125	111	97	83	68	55
12	212	191	173	156	140	126	111	97
14	265	242	223	204	188	171	156	140
16	319	295	275	255	237	219	203	186
18	376	350	327	306	287	269	250	234
20	433	406	382	360	339	320	301	282
22	491	463	437	414	392	371	351	332
24	551	521	494	470	447	425	404	384
26	611	580	551	526	501	478	457	436
28	671	638	610	582	557	534	510	489
30	732	698	668	639	613	589	565	543
32	794	759	727	698	670	645	621	597
34	856	819	787	757	728	702	677	652
36	918	881	847	816	787	759	733	708
38	981	943	908	875	845	817	790	764
40	1044	1004	969	936	905	875	848	821
42	1108	1067	1030	996	964	934	906	879
44	1172	1130	1092	1057	1024	993	964	936
46	1236	1192	1154	1118	1085	1053	1023	994
48	1300	1256	1216	1179	1145	1113	1082	1053
50	1364	1319	1279	1241	1206	1173	1141	1112
52	1429	1383	1342	1303	1267	1233	1201	1171
54	1494	1447	1405	1365	1329	1294	1262	1230
56	1559	1511	1468	1428	1390	1355	1322	1290
58	1625	1576	1532	1491	1452	1416	1383	1350
60	1690	1640	1595	1554	1515	1478	1443	1410
64	1822	1770	1723	1681	1640	1602	1566	1532
68	1954	1901	1853	1808	1766	1727	1689	1654
72	2086	2032	1982	1936	1893	1852	1814	1777
76	2220	2163	2112	2065	2020	1979	1939	1901
80	2353	2295	2243	2194	2148	2105	2065	2025
84	2487	2428	2373	2324	2277	2233	2191	2151
88	2621	2561	2505	2454	2406	2360	2317	2277
92	2756	2694	2637	2585	2535	2489	2445	2403
96	2891	2828	2770	2716	2665	2618	2573	2530
100	3026	2962	2902	2847	2796	2747	2701	2657

Table 3-10. B.01 Peaked Traffic Capacity in CCS (Z = 2.6 to 4.0)

No. of Trunks (N)	Traffic (A) in CCS for Z =							
	2.6	2.8	3.0	3.2	3.4	3.6	3.8	4.0
2								
4								
6								
8								
10	42	28	12					
12	83	69	55	40	26			
14	126	111	97	83	68	53	39	22
16	170	155	141	126	110	96	82	67
18	216	201	185	170	154	140	125	110
20	265	248	231	215	200	184	169	154
22	314	296	279	262	246	230	214	199
24	364	346	328	310	293	277	260	245
26	416	396	378	360	342	325	308	293
28	468	448	428	410	392	373	356	339
30	521	500	480	461	442	423	405	388
32	575	554	532	512	493	473	455	437
34	629	607	586	565	545	525	506	487
36	685	661	639	618	597	576	557	538
38	740	716	694	671	650	629	609	589
40	796	772	748	725	704	682	661	641
42	852	828	804	780	757	736	715	694
44	909	884	859	836	812	790	768	747
46	967	941	916	891	867	845	822	800
48	1025	998	972	947	923	899	876	854
50	1083	1055	1029	1003	979	955	931	908
52	1141	1113	1086	1060	1035	1010	987	963
54	1201	1171	1144	1117	1091	1066	1042	1018
56	1259	1230	1202	1175	1149	1123	1098	1074
58	1319	1289	1260	1233	1206	1180	1154	1130
60	1379	1348	1319	1291	1264	1237	1211	1186
64	1499	1467	1437	1408	1380	1352	1325	1299
68	1620	1587	1556	1526	1498	1468	1440	1414
72	1742	1708	1676	1645	1614	1585	1557	1529
76	1865	1830	1797	1764	1733	1703	1674	1645
80	1988	1953	1918	1885	1853	1822	1792	1763
84	2112	2076	2041	2007	1974	1942	1911	1881
88	2237	2200	2163	2129	2095	2062	2030	1999
92	2363	2324	2287	2252	2217	2183	2151	2119
96	2489	2450	2411	2375	2339	2305	2272	2239
100	2615	2575	2536	2499	2462	2427	2393	2360

Table 3-11. B.02 Peaked Traffic Capacity in CCS (Z = 1.0 to 2.4)

No. of Trunks (N)	Traffic (A) in CCS for Z =							
	1.0	1.2	1.4	1.6	1.8	2.0	2.2	2.4
2	8							
4	39	27	17	7				
6	82	67	55	43	32	20	7	
8	131	114	100	87	74	62	48	36
10	183	165	148	134	120	106	92	80
12	238	219	201	184	168	154	139	125
14	295	274	254	237	220	204	188	173
16	354	330	310	292	273	256	239	223
18	414	389	367	347	328	310	292	275
20	475	449	426	404	384	364	346	328
22	536	509	484	462	440	420	401	382
24	599	570	545	521	498	477	457	437
26	662	632	606	581	557	535	514	494
28	725	694	667	640	616	593	571	550
30	790	758	728	702	676	652	630	608
32	854	821	791	763	737	713	689	666
34	919	885	854	825	798	772	748	725
36	984	949	917	888	869	833	808	784
38	1050	1013	981	950	922	895	869	844
40	1116	1079	1045	1014	984	956	930	905
42	1182	1144	1109	1077	1047	1018	991	965
44	1249	1201	1174	1141	1110	1081	1053	1026
46	1315	1275	1239	1205	1174	1144	1115	1088
48	1382	1341	1304	1270	1237	1206	1177	1150
50	1449	1407	1370	1334	1301	1270	1240	1212
52	1516	1474	1435	1400	1365	1333	1303	1274
54	1584	1540	1501	1464	1430	1397	1366	1337
56	1652	1607	1567	1530	1495	1461	1430	1400
58	1719	1674	1633	1595	1560	1525	1493	1463
60	1787	1741	1700	1661	1625	1590	1557	1526
64	1923	1876	1833	1793	1755	1720	1686	1653
68	2060	2011	1967	1926	1887	1850	1815	1782
72	2197	2147	2102	2059	2019	1981	1945	1911
76	2335	2284	2236	2193	2151	2113	2076	2040
80	2473	2420	2372	2327	2285	2245	2207	2171
84	2542	2557	2508	2462	2418	2378	2339	2302
88	2750	2695	2644	2597	2553	2511	2471	2433
92	2888	2832	2781	2732	2687	2644	2604	2564
96	3028	2970	2917	2868	2822	2778	2737	2697
100	3167	3109	3055	3004	2957	2913	2870	2829

Table 3-12. B.02 Peaked Traffic Capacity in CCS (Z = 2.6 to 4.0)

No. of Trunks (N)	Traffic (A) in CCS for Z =							
	2.6	2.8	3.0	3.2	3.4	3.6	3.8	4.0
2								
4								
6								
8	23	8						
10	66	54	40	27	9			
12	111	98	84	71	58	45	30	12
14	159	144	130	116	103	90	76	62
16	208	193	178	163	150	135	122	109
18	259	243	227	212	198	183	169	155
20	311	294	278	262	247	232	217	203
22	364	347	330	314	297	282	267	251
24	418	401	383	366	349	333	318	301
26	474	455	437	419	402	385	369	353
28	531	510	492	473	456	438	421	404
30	587	567	547	528	510	492	474	457
32	644	624	603	584	565	546	528	511
34	703	681	661	640	620	601	583	564
36	762	739	718	697	677	657	638	619
38	821	798	776	755	734	714	694	675
40	880	857	834	812	791	770	750	730
42	940	916	893	871	849	827	807	787
44	1001	976	953	930	907	885	864	843
46	1061	1037	1012	989	966	944	922	901
48	1123	1097	1072	1048	1025	1002	980	958
50	1184	1158	1133	1108	1085	1061	1039	1016
52	1246	1219	1193	1168	1144	1120	1097	1075
54	1308	1281	1254	1229	1204	1180	1157	1134
56	1370	1343	1316	1290	1265	1240	1216	1193
58	1433	1405	1377	1351	1325	1300	1276	1252
60	1496	1467	1439	1412	1386	1361	1336	1312
64	1622	1592	1564	1536	1509	1483	1457	1432
68	1750	1719	1689	1660	1632	1605	1579	1553
72	1878	1846	1815	1786	1757	1729	1702	1675
76	2006	1974	1942	1912	1882	1854	1826	1799
80	2136	2102	2070	2038	2008	1979	1950	1922
84	2266	2231	2198	2166	2135	2105	2075	2047
88	2396	2361	2327	2294	2262	2231	2201	2172
92	2527	2491	2456	2423	2390	2358	2328	2298
96	2658	2622	2586	2552	2518	2486	2455	2424
100	2790	2753	2716	2681	2647	2614	2582	2551

4

Erlang C Distribution

The Erlang C distribution is used for dimensioning common-equipment server pools where call attempts wait in a first-in, first-out (FIFO) queue until an idle server is available. It is based on the following assumptions:

- Calls are served in order of arrival.
- There are an infinite number of sources.
- Blocked calls are delayed.
- Holding times are exponential.

4.1   ERLANG C FORMULA

The Erlang C Formula given in Equation 4.1, also known as the *Erlang formula of the second kind* $[E_2(N, A)]$, is used to predict the probability that a call will be delayed. Equation 4.2 is used to predict the probability that a call will be delayed longer than a given time. Equations 4.3 and 4.4 are used to predict the average delay on all calls and the average delay on calls delayed, respectively.

$$P(>0) = E_2(N, A) = \cfrac{\cfrac{A^N N}{N! \ (N-1)}}{\displaystyle\sum_{i=0}^{N-1} \frac{A^i}{i!} + \frac{A^N N}{N! \ (N-1)}} \tag{4.1}$$

$$P(>t) = P(>0)\varepsilon^{-(N-A)\frac{T_1}{T_2}} \tag{4.2}$$

$$D_1 = P(>0) \frac{T_2}{(N-A)} \tag{4.3}$$

$$D_2 = \frac{T_2}{(N-A)} \tag{4.4}$$

where $P(>0)$ = Probability of delay greater than zero
 $P(>t)$ = Probability of delay greater than T_1
 N = Number of servers in full-availability group
 A = Traffic offered to group in Erlangs
 T_1 = Acceptable delay time in seconds
 T_2 = Mean server-holding time in seconds
 D_1 = Average delay on all calls in seconds
 D_2 = Average delay on calls delayed in seconds
 ε = Natural logarithm base (2.71828 . . .)

4.2   ERLANG C COMPUTER PROGRAM

The following computer program can be used to calculate Equations 4.1 through 4.4 to determine the Erlang C delay loss probabilities and delays. Required inputs are the number of servers in the group, the traffic offered to the group expressed in Erlangs, the acceptable delay time expressed in seconds, and server-holding time expressed in seconds.

```
100 REM ERLANG C DELAY LOSS CALCULATIONS
110 INPUT "ENTER NUMBER OF SERVERS (N)";N
120 INPUT "ENTER OFFERED TRAFFIC IN ERLANGS (A)";A
130 INPUT "ENTER ACCEPTABLE DELAY (T1)IN SEC.";T1
```

```
140 INPUT "ENTER SERVER-HOLDING TIME (T2)IN SEC.";T2
150 LET X=1
160 LET Y=0
170 FOR I=1 TO N
180 LET Y=X+Y
190 LET X=X*A/I
200 NEXT I
210 LET P0=((N/(N-A))*X)/(Y+((N/(N-A))*X))
220 LET P1=P0*EXP(-(N-A)*T1/T2)
230 PRINT USING "P(>0) = #.#####";P0
240 PRINT USING "P(>T) = #.#####";P1
250 PRINT USING "D1 = ##.#### SEC.";P0*T2/(N-A)
260 PRINT USING "D2 = ##.#### SEC.";T2/(N-A)
270 END
```

4.3   ERLANG C DELAY LOSS PROBABILITY TABLES

Erlang C delay loss probability tables are used to determine the probability that a call will be delayed and the probability that a call will be delayed longer than a given time (expressed as a ratio of delay time to server-holding time). Tables 4-1 through 4-24 present delay loss probabilities for common-equipment groups containing 2 to 25 servers and delay-to-holding time ratios of 0.2 to 2.0. Blanks (no data) in the tables indicate that the delay loss probability is less than 0.000005 (i.e., not necessarily zero, but essentially nonblocking for practical applications). The following examples illustrate typical table usage:

Example 4-1

Determine the delay loss probabilities and delays for a 24-server common-equipment pool if the busy-hour traffic is 20 Erlangs, the acceptable delay time is 4 seconds, and the mean server-holding time (negative exponential distribution) is 4 seconds.

Using Table 4-23 (pages 127–128) (24 servers), read down the $P(>0)$ column to the A row for 20 Erlangs and read .29807 (probability the call will be delayed) at the intersection. Then read down the 1.0 column (4 sec/4 sec) to the A row for 20 Erlangs and read .00546 (probability the call will be delayed longer than 4 seconds) at the intersection.

$$D_1 = P(>0)\left(\frac{T_2}{(N-A)}\right) = \frac{(0.29807)(4\ \text{seconds})}{(24-20)} = 0.29807\ \text{seconds}$$

$$D_2 = \frac{T_2}{(N-A)} = \frac{(4 \text{ seconds})}{(24-20)} = 1 \text{ second}$$

Example 4-2

For the common-equipment pool of Example 4-1, determine the offered traffic in Erlangs and CCS to achieve a grade of service of 0.002 or better.

Using Table 4-23 (24 servers), read down the 1.0 column until .00179 is found. Read across that *A* row to determine that 19.2 Erlangs of traffic can be offered.

$$(19.2 \text{ Erl})(36 \text{ CCS/Erl}) = 691.2 \text{ CCS}$$

Example 4-3

Determine the number of servers required to handle a busy-hour traffic level of 10 Erlangs at a grade of service of 0.001 or better, given that the acceptable delay time is 4 seconds and the mean server-holding time (negative exponential distribution) is 2 seconds.

Using a trial-and-error method, find that Table 4-12 (pages 105–106) (13 servers) yields a grade of service of 0.00071 at the intersection of the 2.0 column and the *A* row for 10 Erlangs. Therefore, at least 13 servers are required.

Table 4-1. Erlang C Delay Loss Probability ($N = 2$ Servers)

Offered Traffic (A in Erl)	P (>0)	Delay Loss Probability P (>t) for T_1/T_2 =				
		0.2	0.4	0.6	0.8	1.0
0.11	.00573	.00393	.00269	.00185	.00126	.00087
0.12	.00679	.00466	.00320	.00220	.00151	.00104
0.13	.00793	.00546	.00376	.00258	.00178	.00122
0.14	.00916	.00631	.00435	.00300	.00207	.00143
0.15	.01047	.00723	.00499	.00345	.00238	.00165
0.16	.01185	.00820	.00568	.00393	.00272	.00188
0.17	.01332	.00924	.00641	.00444	.00308	.00214
0.18	.01486	.01033	.00718	.00499	.00347	.00241
0.19	.01648	.01148	.00799	.00556	.00387	.00270
0.20	.01818	.01269	.00885	.00617	.00431	.00301
0.21	.01995	.01395	.00975	.00682	.00477	.00333
0.22	.02180	.01527	.01070	.00749	.00525	.00368
0.23	.02372	.01665	.01169	.00820	.00576	.00404
0.24	.02571	.01808	.01272	.00894	.00629	.00442
0.25	.02778	.01957	.01379	.00972	.00685	.00483
0.26	.02991	.02112	.01491	.01053	.00744	.00525
0.27	.03211	.02272	.01608	.01137	.00805	.00569
0.28	.03439	.02438	.01728	.01225	.00869	.00616
0.29	.03672	.02609	.01853	.01316	.00935	.00664
0.30	.03913	.02785	.01982	.01411	.01004	.00715
0.31	.04160	.02967	.02116	.01509	.01076	.00768
0.32	.04414	.03154	.02254	.01611	.01151	.00823
0.33	.04674	.03347	.02396	.01716	.01229	.00880
0.34	.04940	.03545	.02543	.01825	.01309	.00939
0.35	.05213	.03748	.02694	.01937	.01393	.01001
0.36	.05492	.03956	.02850	.02053	.01479	.01065
0.37	.05776	.04169	.03010	.02172	.01568	.01132
0.38	.06067	.04388	.03174	.02295	.01660	.01201
0.39	.06364	.04612	.03342	.02422	.01755	.01272
0.40	.06667	.04841	.03515	.02553	.01854	.01346
0.41	.06975	.05075	.03693	.02687	.01955	.01422
0.42	.07289	.05314	.03874	.02825	.02059	.01501
0.43	.07609	.05559	.04061	.02966	.02167	.01583
0.44	.07934	.05808	.04251	.03112	.02278	.01667
0.45	.08265	.06062	.04446	.03261	.02392	.01754
0.46	.08602	.06321	.04646	.03414	.02509	.01844
0.47	.08943	.06586	.04850	.03571	.02630	.01937
0.48	.09290	.06855	.05058	.03732	.02754	.02032
0.49	.09643	.07129	.05271	.03897	.02881	.02130
0.50	.10000	.07408	.05488	.04066	.03012	.02231

(table continues)

Table 4-1.   Erlang C Delay Loss Probability ($N = 2$ Servers *Continued*)

Offered Traffic (*A* in Erl)	$P(>0)$	Delay Loss Probability $P(>t)$ for $T_1/T_2 =$				
		1.2	1.4	1.6	1.8	2.0
0.11	.00573	.00059	.00041	.00028	.00019	.00013
0.12	.00679	.00071	.00049	.00034	.00023	.00016
0.13	.00793	.00084	.00058	.00040	.00027	.00019
0.14	.00916	.00098	.00068	.00047	.00032	.00022
0.15	.01047	.00114	.00079	.00054	.00037	.00026
0.16	.01185	.00130	.00090	.00062	.00043	.00030
0.17	.01332	.00148	.00103	.00071	.00049	.00034
0.18	.01486	.00167	.00116	.00081	.00056	.00039
0.19	.01648	.00188	.00131	.00091	.00063	.00044
0.20	.01818	.00210	.00146	.00102	.00071	.00050
0.21	.01995	.00233	.00163	.00114	.00080	.00056
0.22	.02180	.00258	.00180	.00126	.00089	.00062
0.23	.02372	.00284	.00199	.00140	.00098	.00069
0.24	.02571	.00311	.00219	.00154	.00108	.00076
0.25	.02778	.00340	.00240	.00169	.00119	.00084
0.26	.02991	.00371	.00262	.00185	.00131	.00092
0.27	.03211	.00403	.00285	.00202	.00143	.00101
0.28	.03439	.00437	.00309	.00219	.00156	.00110
0.29	.03672	.00472	.00335	.00238	.00169	.00120
0.30	.03913	.00509	.00362	.00258	.00183	.00131
0.31	.04160	.00547	.00390	.00278	.00199	.00142
0.32	.04414	.00588	.00420	.00300	.00215	.00153
0.33	.04674	.00630	.00451	.00323	.00231	.00166
0.34	.04940	.00674	.00484	.00347	.00249	.00179
0.35	.05213	.00720	.00517	.00372	.00267	.00192
0.36	.05492	.00767	.00553	.00398	.00287	.00207
0.37	.05776	.00817	.00590	.00426	.00307	.00222
0.38	.06067	.00868	.00628	.00454	.00329	.00238
0.39	.06364	.00922	.00668	.00484	.00351	.00254
0.40	.06667	.00977	.00710	.00515	.00374	.00272
0.41	.06975	.01035	.00753	.00548	.00399	.00290
0.42	.07289	.01095	.00798	.00582	.00424	.00309
0.43	.07609	.01156	.00845	.00617	.00451	.00329
0.44	.07934	.01220	.00893	.00654	.00479	.00350
0.45	.08265	.01287	.00944	.00692	.00508	.00372
0.46	.08602	.01355	.00996	.00732	.00538	.00395
0.47	.08943	.01426	.01050	.00773	.00569	.00419
0.48	.09290	.01499	.01106	.00816	.00602	.00444
0.49	.09643	.01575	.01164	.00861	.00636	.00471
0.50	.10000	.01653	.01225	.00907	.00672	.00498

Table 4-2. Erlang C Delay Loss Probability ($N = 3$ Servers)

Offered Traffic (A in Erl)	$P(>0)$	Delay Loss Probability $P(>t)$ for $T_1/T_2 =$				
		0.2	0.4	0.6	0.8	1.0
0.13	.00034	.00019	.00011	.00006	.00003	.00002
0.16	.00061	.00035	.00020	.00011	.00006	.00003
0.19	.00101	.00058	.00033	.00019	.00011	.00006
0.22	.00154	.00088	.00051	.00029	.00017	.00010
0.25	.00221	.00128	.00074	.00042	.00025	.00014
0.28	.00305	.00177	.00103	.00060	.00035	.00020
0.31	.00406	.00237	.00138	.00081	.00047	.00028
0.34	.00526	.00309	.00181	.00107	.00063	.00037
0.37	.00665	.00393	.00232	.00137	.00081	.00048
0.40	.00825	.00490	.00292	.00173	.00103	.00061
0.43	.01006	.00602	.00360	.00215	.00129	.00077
0.46	.01209	.00727	.00438	.00263	.00158	.00095
0.49	.01435	.00868	.00526	.00318	.00193	.00117
0.52	.01684	.01025	.00624	.00380	.00232	.00141
0.55	.01957	.01199	.00734	.00450	.00276	.00169
0.58	.02254	.01389	.00856	.00528	.00325	.00200
0.61	.02576	.01597	.00990	.00614	.00381	.00236
0.64	.02923	.01823	.01137	.00709	.00442	.00276
0.67	.03295	.02067	.01297	.00814	.00511	.00321
0.70	.03692	.02331	.01471	.00929	.00586	.00370
0.73	.04115	.02614	.01660	.01054	.00669	.00425
0.76	.04564	.02916	.01863	.01190	.00761	.00486
0.79	.05039	.03239	.02082	.01338	.00860	.00553
0.82	.05540	.03582	.02316	.01498	.00969	.00626
0.85	.06067	.03947	.02567	.01670	.01086	.00707
0.88	.06620	.04332	.02835	.01855	.01214	.00795
0.91	.07199	.04739	.03120	.02054	.01352	.00890
0.94	.07804	.05169	.03423	.02267	.01502	.00995
0.97	.08434	.05620	.03745	.02495	.01662	.01108
1.00	.09091	.06094	.04085	.02738	.01835	.01230
1.03	.09773	.06591	.04444	.02997	.02021	.01363
1.06	.10481	.07110	.04824	.03273	.02220	.01506
1.09	.11214	.07654	.05224	.03565	.02433	.01661
1.12	.11973	.08220	.05644	.03875	.02661	.01927
1.15	.12756	.08811	.06086	.04204	.02904	.02006
1.18	.13565	.09426	.06550	.04552	.03163	.02198
1.21	.14398	.10065	.07036	.04919	.03439	.02404
1.24	.15256	.10729	.07546	.05307	.03732	.02625
1.27	.16138	.11418	.08078	.05715	.04044	.02861
1.30	.17044	.12132	.08635	.06146	.04375	.03114

(table continues)

Table 4-2. Erlang C Delay Loss Probability (N = 3 Servers *Continued*)

Offered Traffic (A in Erl)	P (>0)	Delay Loss Probability P (>t) for T_1/T_2 =				
		1.2	1.4	1.6	1.8	2.0
0.13	.00034	.00001	.00001			
0.16	.00061	.00002	.00001	.00001		
0.19	.00101	.00003	.00002	.00001	.00001	
0.22	.00154	.00005	.00003	.00002	.00001	.00001
0.25	.00221	.00008	.00005	.00003	.00002	.00001
0.28	.00305	.00012	.00007	.00004	.00002	.00001
0.31	.00406	.00016	.00009	.00005	.00003	.00002
0.34	.00526	.00022	.00013	.00007	.00004	.00003
0.37	.00665	.00028	.00017	.00010	.00006	.00003
0.40	.00825	.00036	.00022	.00013	.00008	.00005
0.43	.01006	.00046	.00028	.00016	.00010	.00006
0.46	.01209	.00057	.00035	.00021	.00012	.00008
0.49	.01435	.00071	.00043	.00026	.00016	.00009
0.52	.01684	.00086	.00052	.00032	.00019	.00012
0.55	.01957	.00103	.00063	.00039	.00024	.00015
0.58	.02254	.00124	.00076	.00047	.00029	.00018
0.61	.02576	.00146	.00091	.00056	.00035	.00022
0.64	.02923	.00172	.00107	.00067	.00042	.00026
0.67	.03295	.00201	.00126	.00079	.00050	.00031
0.70	.03692	.00234	.00148	.00093	.00059	.00037
0.73	.04115	.00270	.00171	.00109	.00069	.00044
0.76	.04564	.00310	.00198	.00127	.00081	.00052
0.79	.05039	.00355	.00228	.00147	.00094	.00061
0.82	.05540	.00405	.00262	.00169	.00109	.00071
0.85	.06067	.00460	.00299	.00195	.00127	.00082
0.88	.06620	.00520	.00340	.00223	.00146	.00095
0.91	.07199	.00586	.00386	.00254	.00167	.00110
0.94	.07804	.00659	.00436	.00289	.00191	.00127
0.97	.08434	.00738	.00492	.00328	.00218	.00145
1.00	.09091	.00825	.00553	.00371	.00248	.00167
1.03	.09773	.00919	.00620	.00418	.00282	.00190
1.06	.10481	.01022	.00693	.00470	.00319	.00216
1.09	.11214	.01133	.00774	.00528	.00360	.00246
1.12	.11973	.01254	.00861	.00591	.00406	.00279
1.15	.12756	.01385	.00957	.00661	.00457	.00315
1.18	.13565	.01527	.01061	.00737	.00512	.00356
1.21	.14398	.01681	.01175	.00821	.00574	.00401
1.24	.15256	.01846	.01298	.00913	.00642	.00452
1.27	.16138	.02024	.01432	.01013	.00717	.00507
1.30	.17044	.02216	.01577	.01123	.00799	.00569

Table 4-3. Erlang C Delay Loss Probability ($N = 4$ Servers)

Offered Traffic (A in Erl)	$P(>0)$	Delay Loss Probability $P(>t)$ for $T_1/T_2 =$				
		0.2	0.4	0.6	0.8	1.0
0.54	.00239	.00119	.00060	.00030	.00015	.00008
0.58	.00309	.00156	.00079	.00040	.00020	.00010
0.62	.00392	.00199	.00101	.00052	.00026	.00013
0.66	.00489	.00251	.00129	.00066	.00034	.00017
0.70	.00602	.00311	.00161	.00083	.00043	.00022
0.74	.00731	.00381	.00198	.00103	.00054	.00028
0.78	.00878	.00461	.00242	.00127	.00067	.00035
0.82	.01043	.00552	.00292	.00155	.00082	.00043
0.86	.01228	.00655	.00350	.00187	.00100	.00053
0.90	.01433	.00771	.00415	.00223	.00120	.00065
0.94	.01659	.00900	.00488	.00265	.00143	.00078
0.98	.01908	.01043	.00570	.00312	.00170	.00093
1.02	.02180	.01201	.00662	.00365	.00201	.00111
1.06	.02475	.01375	.00764	.00424	.00236	.00131
1.10	.02795	.01565	.00876	.00491	.00275	.00154
1.14	.03139	.01772	.01000	.00564	.00319	.00180
1.18	.03510	.01977	.01136	.00646	.00368	.00209
1.22	.03907	.02240	.01285	.00737	.00423	.00242
1.26	.04330	.02503	.01447	.00837	.00484	.00280
1.30	.04781	.02786	.01624	.00946	.00551	.00321
1.34	.05260	.03090	.01815	.01066	.00626	.00368
1.38	.05766	.03415	.02022	.01197	.00709	.00420
1.42	.06302	.03761	.02245	.01340	.00800	.00477
1.46	.06866	.04131	.02486	.01496	.00900	.00541
1.50	.07459	.04524	.02744	.01664	.01009	.00612
1.54	.08081	.04941	.03021	.01847	.01129	.00690
1.58	.08733	.05382	.03317	.02044	.01260	.00777
1.62	.09415	.05849	.03634	.02257	.01402	.00871
1.66	.10126	.06341	.03971	.02487	.01558	.00975
1.70	.10868	.06861	.04331	.02734	.01726	.01090
1.74	.11639	.07407	.04713	.02999	.01909	.01215
1.78	.12441	.07981	.05119	.03284	.02106	.01351
1.82	.13273	.08583	.05550	.03589	.02320	.01500
1.86	.14135	.09214	.06006	.03914	.02552	.01663
1.90	.15028	.09874	.06488	.04263	.02801	.01840
1.94	.15951	.10565	.06997	.04634	.03069	.02033
1.98	.16904	.11286	.07535	.05031	.03359	.02242
2.02	.17887	.12038	.08101	.05452	.03669	.02470
2.06	.18900	.12822	.08698	.05901	.04003	.02716
2.10	.19942	.13638	.09326	.06378	.04362	.02983

(*table continues*)

Table 4-3. Erlang C Delay Loss Probability ($N = 4$ Servers *Continued*)

Offered Traffic (A in Erl)	$P(>0)$	Delay Loss Probability $P(>t)$ for $T_1/T_2 =$				
		1.2	1.4	1.6	1.8	2.0
0.54	.00239	.00004	.00002	.00001		
0.58	.00309	.00005	.00003	.00001	.00001	
0.62	.00392	.00007	.00003	.00002	.00001	
0.66	.00489	.00009	.00005	.00002	.00001	.00001
0.70	.00602	.00011	.00006	.00003	.00002	.00001
0.74	.00731	.00015	.00008	.00004	.00002	.00001
0.78	.00878	.00018	.00010	.00005	.00003	.00001
0.82	.01043	.00023	.00012	.00006	.00003	.00002
0.86	.01228	.00028	.00015	.00008	.00004	.00002
0.90	.01433	.00035	.00019	.00010	.00005	.00003
0.94	.01659	.00042	.00023	.00012	.00007	.00004
0.98	.01908	.00051	.00028	.00015	.00008	.00005
1.02	.02180	.00061	.00034	.00019	.00010	.00006
1.06	.02475	.00073	.00040	.00022	.00012	.00007
1.10	.02795	.00086	.00048	.00027	.00015	.00008
1.14	.03139	.00101	.00057	.00032	.00018	.00010
1.18	.03510	.00119	.00068	.00039	.00022	.00012
1.22	.03907	.00139	.00080	.00046	.00026	.00015
1.26	.04330	.00162	.00093	.00054	.00031	.00018
1.30	.04781	.00187	.00109	.00064	.00037	.00022
1.34	.05260	.00216	.00127	.00075	.00044	.00026
1.38	.05766	.00249	.00147	.00087	.00052	.00031
1.42	.06302	.00285	.00170	.00102	.00061	.00036
1.46	.06866	.00326	.00196	.00118	.00071	.00043
1.50	.07459	.00371	.00225	.00137	.00083	.00050
1.54	.08081	.00422	.00258	.00158	.00096	.00059
1.58	.08733	.00479	.00295	.00182	.00112	.00069
1.62	.09415	.00541	.00336	.00209	.00130	.00081
1.66	.10126	.00611	.00383	.00240	.00150	.00094
1.70	.10868	.00688	.00434	.00274	.00173	.00109
1.74	.11639	.00773	.00492	.00313	.00199	.00127
1.78	.12441	.00867	.00556	.00357	.00229	.00147
1.82	.13273	.00970	.00627	.00406	.00262	.00170
1.86	.14135	.01084	.00707	.00461	.00300	.00196
1.90	.15028	.01209	.00794	.00522	.00343	.00225
1.94	.15951	.01346	.00892	.00591	.00391	.00259
1.98	.16904	.01497	.01000	.00667	.00446	.00297
2.02	.17887	.01662	.01119	.00753	.00507	.00341
2.06	.18900	.01843	.01250	.00848	.00575	.00390
2.10	.19942	.02040	.01395	.00954	.00652	.00446

Table 4-4. Erlang C Delay Loss Probability ($N = 5$ Servers)

Offered Traffic (A in Erl)	$P(>0)$	Delay Loss Probability $P(>t)$ for $T_1/T_2 =$				
		0.2	0.4	0.6	0.8	1.0
1.05	.00471	.00214	.00097	.00044	.00020	.00009
1.10	.00573	.00262	.00120	.00055	.00025	.00012
1.15	.00689	.00319	.00148	.00068	.00032	.00015
1.20	.00821	.00384	.00180	.00084	.00039	.00018
1.25	.00971	.00459	.00217	.00102	.00048	.00023
1.30	.01139	.00543	.00259	.00124	.00059	.00028
1.35	.01326	.00639	.00308	.00148	.00072	.00034
1.40	.01533	.00746	.00363	.00177	.00086	.00042
1.45	.01762	.00866	.00426	.00209	.00103	.00051
1.50	.02014	.01000	.00497	.00247	.00122	.00061
1.55	.02289	.01148	.00576	.00289	.00145	.00073
1.60	.02589	.01311	.00664	.00337	.00171	.00086
1.65	.02914	.01491	.00763	.00390	.00200	.00102
1.70	.03265	.01687	.00872	.00451	.00233	.00120
1.75	.03643	.01902	.00993	.00518	.00271	.00141
1.80	.04050	.02135	.01126	.00594	.00313	.00165
1.85	.04485	.02389	.01272	.00678	.00361	.00192
1.90	.04950	.02663	.01432	.00771	.00414	.00223
1.95	.05444	.02958	.01607	.00873	.00475	.00258
2.00	.05970	.03276	.01798	.00987	.00542	.00297
2.05	.06527	.03618	.02006	.01112	.00616	.00342
2.10	.07116	.03984	.02231	.01249	.00699	.00392
2.15	.07738	.04376	.02475	.01400	.00791	.00448
2.20	.08393	.04794	.02738	.01564	.00893	.00510
2.25	.09081	.05239	.03023	.01744	.01006	.00581
2.30	.09803	.05713	.03329	.01940	.01131	.00659
2.35	.10559	.06215	.03658	.02153	.01267	.00746
2.40	.11350	.06748	.04012	.02385	.01418	.00843
2.45	.12176	.07312	.04391	.02637	.01583	.00951
2.50	.13037	.07907	.04796	.02909	.01764	.01070
2.55	.13993	.08536	.05229	.03204	.01963	.01202
2.60	.14865	.09198	.05692	.03522	.02179	.01349
2.65	.15833	.09896	.06185	.03866	.02416	.01510
2.70	.16837	.10629	.06710	.04236	.02674	.01688
2.75	.17876	.11398	.07268	.04634	.02955	.01884
2.80	.18952	.12206	.07861	.05063	.03261	.02100
2.85	.20063	.13051	.08490	.05523	.03593	.02337
2.90	.21211	.13937	.09157	.06017	.03953	.02597
2.95	.22393	.14862	.09863	.06546	.04344	.02883
3.00	.23615	.15830	.10611	.07113	.04768	.03196

(table continues)

Table 4-4. Erlang C Delay Loss Probability (*N* = 5 Servers *Continued*)

Offered Traffic (*A* in Erl)	*P* (>0)	Delay Loss Probability *P* (>*t*) for T_1/T_2 =				
		1.2	1.4	1.6	1.8	2.0
1.05	.00471	.00004	.00002	.00001		
1.10	.00573	.00005	.00002	.00001	.00001	
1.15	.00689	.00007	.00003	.00001	.00001	
1.20	.00821	.00009	.00004	.00002	.00001	
1.25	.00971	.00011	.00005	.00002	.00001	.00001
1.30	.01139	.00013	.00006	.00003	.00001	.00001
1.35	.01326	.00017	.00008	.00004	.00002	.00001
1.40	.01533	.00020	.00010	.00005	.00002	.00001
1.45	.01762	.00025	.00012	.00006	.00003	.00001
1.50	.02014	.00030	.00015	.00007	.00004	.00002
1.55	.02289	.00036	.00018	.00009	.00005	.00002
1.60	.02589	.00044	.00022	.00011	.00006	.00003
1.65	.02914	.00052	.00027	.00014	.00007	.00004
1.70	.03265	.00062	.00032	.00017	.00009	.00004
1.75	.03643	.00074	.00038	.00020	.00010	.00005
1.80	.04050	.00087	.00046	.00024	.00013	.00007
1.85	.04485	.00102	.00055	.00029	.00015	.00008
1.90	.04950	.00120	.00065	.00035	.00019	.00010
1.95	.05444	.00140	.00076	.00041	.00022	.00012
2.00	.05970	.00163	.00090	.00049	.00027	.00015
2.05	.06527	.00189	.00105	.00058	.00032	.00018
2.10	.07116	.00219	.00123	.00069	.00038	.00022
2.15	.07738	.00253	.00143	.00081	.00046	.00026
2.20	.08393	.00292	.00167	.00095	.00054	.00031
2.25	.09081	.00335	.00193	.00111	.00064	.00037
2.30	.09803	.00384	.00224	.00130	.00076	.00044
2.35	.10559	.00439	.00258	.00152	.00090	.00053
2.40	.11350	.00501	.00298	.00177	.00105	.00063
2.45	.12176	.00571	.00343	.00206	.00124	.00074
2.50	.13037	.00649	.00394	.00239	.00145	.00088
2.55	.13993	.00737	.00451	.00276	.00169	.00104
2.60	.14865	.00834	.00516	.00320	.00198	.00122
2.65	.15833	.00944	.00590	.00369	.00230	.00144
2.70	.16837	.01066	.00673	.00425	.00268	.00169
2.75	.17876	.01201	.00766	.00488	.00311	.00199
2.80	.18952	.01352	.00871	.00561	.00361	.00233
2.85	.20063	.01520	.00989	.00643	.00418	.00272
2.90	.21211	.01707	.01121	.00737	.00484	.00318
2.95	.22393	.01913	.01270	.00843	.00559	.00371
3.00	.23615	.02142	.01436	.00963	.00645	.00433

Table 4-5. Erlang C Delay Loss Probability (N = 6 Servers)

Offered Traffic (A in Erl)	P (>0)	Delay Loss Probability P (>t) for T_1/T_2 =				
		0.2	0.4	0.6	0.8	1.0
1.80	.01115	.00481	.00208	.00090	.00039	.00017
1.85	.01265	.00551	.00240	.00105	.00046	.00020
1.90	.01429	.00629	.00277	.00122	.00054	.00024
1.95	.01607	.00715	.00318	.00142	.00063	.00028
2.00	.01802	.00810	.00364	.00163	.00073	.00033
2.05	.02012	.00913	.00414	.00188	.00085	.00039
2.10	.02240	.01027	.00471	.00216	.00099	.00045
2.15	.02485	.01150	.00533	.00247	.00114	.00053
2.20	.02748	.01285	.00601	.00281	.00131	.00061
2.25	.03030	.01431	.00676	.00319	.00151	.00071
2.30	.03331	.01589	.00758	.00362	.00173	.00082
2.35	.03653	.01760	.00848	.00409	.00197	.00095
2.40	.03995	.01945	.00947	.00461	.00224	.00109
2.45	.04359	.02143	.01054	.00518	.00255	.00125
2.50	.04744	.02356	.01170	.00581	.00289	.00143
2.55	.05152	.02584	.01296	.00650	.00326	.00164
2.60	.05583	.02829	.01433	.00726	.00368	.00186
2.65	.06038	.03090	.01581	.00809	.00414	.00212
2.70	.06516	.03368	.01741	.00900	.00465	.00240
2.75	.07019	.03664	.01913	.00999	.00521	.00272
2.80	.07547	.03979	.02098	.01106	.00583	.00308
2.85	.08100	.04314	.02297	.01224	.00652	.00347
2.90	.08678	.04668	.02511	.01351	.00727	.00391
2.95	.09283	.05044	.02741	.01489	.00809	.00440
3.00	.09914	.05441	.02986	.01639	.00899	.00494
3.05	.10572	.05861	.03249	.01801	.00998	.00553
3.10	.11257	.06303	.03529	.01976	.01106	.00619
3.15	.11970	.06769	.03828	.02165	.01224	.00692
3.20	.12710	.07260	.04147	.02369	.01353	.00773
3.25	.13479	.07776	.04487	.02589	.01493	.00862
3.30	.14275	.08319	.04848	.02825	.01646	.00959
3.35	.15100	.08888	.05231	.03079	.01812	.01067
3.40	.15953	.09485	.05639	.03352	.01993	.01185
3.45	.16835	.10110	.06071	.03654	.02189	.01315
3.50	.17747	.10764	.06529	.03960	.02402	.01457
3.55	.18687	.11448	.07013	.04297	.02632	.01613
3.60	.19657	.12163	.07526	.04657	.02882	.01783
3.65	.20656	.12910	.08069	.05043	.03152	.01970
3.70	.21684	.13689	.08641	.05455	.03444	.02174
3.75	.22742	.14501	.09246	.05896	.03759	.02397

(table continues)

Table 4-5. Erlang C Delay Loss Probability (*N* = 6 Servers *Continued*)

Offered Traffic (*A* in Erl)	*P* (>0)	Delay Loss Probability *P* (>*t*) for T_1/T_2 =				
		1.2	1.4	1.6	1.8	2.0
1.80	.01115	.00007	.00003	.00001	.00001	
1.85	.01265	.00009	.00004	.00002	.00001	
1.90	.01429	.00010	.00005	.00002	.00001	
1.95	.01607	.00012	.00006	.00002	.00001	
2.00	.01802	.00015	.00007	.00003	.00001	.00001
2.05	.02012	.00018	.00008	.00004	.00002	.00001
2.10	.02240	.00021	.00010	.00004	.00002	.00001
2.15	.02485	.00024	.00011	.00005	.00002	.00001
2.20	.02748	.00029	.00013	.00006	.00003	.00001
2.25	.03030	.00034	.00016	.00008	.00004	.00002
2.30	.03331	.00039	.00019	.00009	.00004	.00002
2.35	.03653	.00046	.00022	.00011	.00005	.00002
2.40	.03995	.00053	.00026	.00013	.00006	.00003
2.45	.04359	.00062	.00030	.00015	.00007	.00004
2.50	.04744	.00071	.00035	.00018	.00009	.00004
2.55	.05152	.00082	.00041	.00021	.00010	.00005
2.60	.05583	.00094	.00048	.00024	.00012	.00006
2.65	.06038	.00108	.00055	.00028	.00015	.00007
2.70	.06516	.00124	.00064	.00033	.00017	.00009
2.75	.07019	.00142	.00074	.00039	.00020	.00011
2.80	.07547	.00162	.00086	.00045	.00024	.00013
2.85	.08100	.00185	.00098	.00052	.00028	.00015
2.90	.08678	.00210	.00113	.00061	.00033	.00018
2.95	.09283	.00239	.00130	.00071	.00038	.00021
3.00	.09914	.00271	.00149	.00082	.00045	.00025
3.05	.10572	.00307	.00170	.00094	.00052	.00029
3.10	.11257	.00347	.00194	.00109	.00061	.00034
3.15	.11970	.00392	.00221	.00125	.00071	.00040
3.20	.12710	.00441	.00252	.00144	.00082	.00047
3.25	.13479	.00497	.00287	.00165	.00095	.00055
3.30	.14275	.00559	.00326	.00190	.00111	.00064
3.35	.15100	.00628	.00370	.00218	.00128	.00075
3.40	.15953	.00704	.00419	.00249	.00148	.00088
3.45	.16835	.00789	.00474	.00285	.00171	.00103
3.50	.17747	.00884	.00536	.00325	.00197	.00120
3.55	.18687	.00988	.00605	.00371	.00227	.00139
3.60	.19657	.01103	.00683	.00422	.00261	.00162
3.65	.20656	.01231	.00769	.00481	.00301	.00188
3.70	.21684	.01372	.00866	.00547	.00345	.00218
3.75	.22742	.01528	.00975	.00621	.00396	.00253

Table 4-6. Erlang C Delay Loss Probability (N = 7 Servers)

Offered Traffic (A in Erl)	P (>0)	Delay Loss Probability P (>t) for T_1/T_2 =				
		0.2	0.4	0.6	0.8	1.0
1.88	.00344	.00123	.00044	.00016	.00006	.00002
1.96	.00431	.00157	.00057	.00021	.00008	.00003
2.04	.00535	.00198	.00074	.00027	.00010	.00004
2.12	.00657	.00248	.00093	.00035	.00013	.00005
2.20	.00799	.00306	.00117	.00045	.00017	.00007
2.28	.00963	.00375	.00146	.00057	.00022	.00009
2.36	.01151	.00455	.00180	.00071	.00028	.00011
2.44	.01365	.00549	.00220	.00089	.00036	.00014
2.52	.01608	.00656	.00268	.00109	.00045	.00018
2.60	.01880	.00780	.00323	.00134	.00056	.00023
2.68	.02184	.00921	.00388	.00164	.00069	.00029
2.74	.02523	.01081	.00463	.00198	.00085	.00036
2.86	.02898	.01261	.00549	.00239	.00104	.00045
2.92	.03311	.01464	.00647	.00286	.00127	.00056
3.00	.03765	.01692	.00760	.00342	.00153	.00069
3.08	.04260	.01945	.00888	.00405	.00185	.00085
3.16	.04799	.02226	.01033	.00479	.00222	.00103
3.24	.05384	.02538	.01197	.00564	.00266	.00125
3.32	.06016	.02882	.01380	.00661	.00317	.00152
3.40	.06697	.03260	.01587	.00772	.00376	.00183
3.48	.07429	.03674	.01817	.00899	.00445	.00220
3.56	.08213	.04127	.02074	.01043	.00524	.00263
3.64	.09050	.04622	.02360	.01205	.00616	.00314
3.72	.09941	.05159	.02677	.01389	.00721	.00374
3.80	.10889	.05742	.03028	.01596	.00842	.00444
3.88	.11894	.06373	.03415	.01830	.00980	.00525
3.96	.12957	.07054	.03841	.02091	.01138	.00620
4.04	.14080	.07789	.04309	.02384	.01319	.00730
4.12	.15262	.08579	.04823	.02711	.01524	.00857
4.20	.16505	.09428	.05385	.03076	.01757	.01004
4.28	.17810	.10337	.06000	.03483	.02021	.01173
4.36	.19178	.11311	.06671	.03934	.02320	.01369
4.44	.20608	.12350	.07401	.04436	.02658	.01593
4.52	.22101	.13459	.08196	.04991	.03039	.01851
4.60	.23659	.14640	.09059	.05605	.03469	.02146
4.68	.25281	.15896	.09995	.06284	.03951	.02484
4.76	.26967	.17229	.11008	.07033	.04493	.02871
4.84	.28718	.18644	.12104	.07858	.05101	.03312
4.92	.30534	.20143	.13288	.08766	.05782	.03815
5.00	.32415	.21728	.14565	.09763	.06544	.04387

(table continues)

Table 4-6. Erlang C Delay Loss Probability ($N = 7$ Servers *Continued*)

Offered Traffic (A in Erl)	$P (>0)$	Delay Loss Probability $P (>t)$ for $T_1/T_2 =$				
		1.2	1.4	1.6	1.8	2.0
1.88	.00344	.00001				
1.96	.00431	.00001				
2.04	.00535	.00001	.00001			
2.12	.00657	.00002	.00001			
2.20	.00799	.00003	.00001			
2.28	.00963	.00003	.00001	.00001		
2.36	.01151	.00004	.00002	.00001		
2.44	.01365	.00006	.00002	.00001		
2.52	.01608	.00007	.00003	.00001	.00001	
2.60	.01880	.00010	.00004	.00002	.00001	
2.68	.02184	.00012	.00005	.00002	.00001	
2.74	.02523	.00016	.00007	.00003	.00001	.00001
2.86	.02898	.00020	.00009	.00004	.00002	.00001
2.92	.03311	.00025	.00011	.00005	.00002	.00001
3.00	.03765	.00031	.00014	.00006	.00003	.00001
3.08	.04260	.00039	.00018	.00008	.00004	.00002
3.16	.04799	.00048	.00022	.00010	.00005	.00002
3.24	.05384	.00059	.00028	.00013	.00006	.00003
3.32	.06016	.00073	.00035	.00017	.00008	.00004
3.40	.06697	.00089	.00043	.00021	.00010	.00005
3.48	.07429	.00109	.00054	.00027	.00013	.00007
3.56	.08213	.00132	.00067	.00033	.00017	.00008
3.64	.09050	.00161	.00082	.00042	.00021	.00011
3.72	.09941	.00194	.00101	.00052	.00027	.00014
3.80	.10889	.00234	.00123	.00065	.00034	.00018
3.88	.11894	.00281	.00151	.00081	.00043	.00023
3.96	.12957	.00337	.00184	.00100	.00054	.00030
4.04	.14080	.00404	.00223	.00124	.00068	.00038
4.12	.15262	.00482	.00271	.00152	.00086	.00048
4.20	.16505	.00573	.00327	.00187	.00107	.00061
4.28	.17810	.00681	.00395	.00229	.00133	.00077
4.36	.19178	.00807	.00476	.00281	.00166	.00098
4.44	.20608	.00955	.00572	.00343	.00205	.00123
4.52	.22101	.01127	.00686	.00418	.00255	.00155
4.60	.23659	.01328	.00822	.00509	.00315	.00195
4.68	.25281	.01562	.00982	.00618	.00388	.00244
4.76	.26967	.01834	.01172	.00749	.00478	.00306
4.84	.28718	.02150	.01396	.00906	.00588	.00382
4.92	.30534	.02516	.01660	.01095	.00722	.00477
5.00	.32415	.02941	.01971	.01321	.00886	.00594

Table 4-7. Erlang C Delay Loss Probability ($N = 8$ Servers)

Offered Traffic (A in Erl)	$P(>0)$	Delay Loss Probability $P(>t)$ for $T_1/T_2 =$				
		0.2	0.4	0.6	0.8	1.0
2.1	.00156	.00048	.00015	.00005	.00001	
2.2	.00208	.00065	.00020	.00006	.00002	.00001
2.3	.00273	.00087	.00028	.00009	.00003	.00001
2.4	.00354	.00115	.00038	.00012	.00004	.00001
2.5	.00452	.00150	.00050	.00017	.00006	.00002
2.6	.00570	.00193	.00066	.00022	.00008	.00003
2.7	.00710	.00246	.00085	.00030	.00010	.00004
2.8	.00876	.00310	.00109	.00039	.00014	.00005
2.9	.01070	.00386	.00139	.00050	.00018	.00007
3.0	.01295	.00476	.00175	.00064	.00024	.00009
3.1	.01554	.00583	.00219	.00082	.00031	.00012
3.2	.01849	.00708	.00271	.00104	.00040	.00015
3.3	.02185	.00854	.00333	.00130	.00051	.00020
3.4	.02564	.01022	.00407	.00162	.00065	.00026
3.5	.02989	.01215	.00494	.00201	.00082	.00033
3.6	.03462	.01436	.00596	.00247	.00102	.00043
3.7	.03988	.01687	.00714	.00302	.00128	.00054
3.8	.04568	.01972	.00851	.00368	.00159	.00069
3.9	.05206	.02293	.01010	.00445	.00196	.00086
4.0	.05904	.02653	.01192	.00536	.00241	.00108
4.1	.06666	.03056	.01401	.00642	.00294	.00135
4.2	.07492	.03504	.01639	.00766	.00358	.00168
4.3	.08387	.04002	.01909	.00911	.00435	.00207
4.4	.09352	.04452	.02216	.01078	.00525	.00256
4.5	.10389	.05159	.02652	.01272	.00632	.00314
4.6	.11500	.05826	.02952	.01495	.00758	.00384
4.7	.12688	.06558	.03389	.01752	.00905	.00468
4.8	.13954	.07358	.03880	.02046	.01079	.00569
4.9	.15300	.08230	.04428	.02382	.01281	.00689
5.0	.16727	.09180	.05038	.02765	.01517	.00833
5.1	.18236	.10210	.05717	.03201	.01792	.01003
5.2	.19829	.11327	.06470	.03696	.02111	.01206
5.3	.21507	.12533	.07304	.04256	.02480	.01445
5.4	.23271	.13835	.08225	.04890	.02907	.01728
5.5	.25122	.15237	.09242	.05605	.03400	.02062
5.6	.27060	.16744	.10361	.06411	.03967	.02455
5.7	.29086	.18362	.11592	.07318	.04619	.02916
5.8	.31201	.20095	.12942	.08335	.05368	.03457
5.9	.33405	.21949	.14421	.09475	.06226	.04091
6.0	.35698	.23929	.16040	.10752	.07207	.04831

(table continues)

Table 4-7. Erlang C Delay Loss Probability ($N = 8$ Servers *Continued*)

Offered Traffic (A in Erl)	$P(>0)$	Delay Loss Probability $P(>t)$ for $T_1/T_2 =$				
		1.2	1.4	1.6	1.8	2.0
2.1	.00156					
2.2	.00208					
2.3	.00273					
2.4	.00354					
2.5	.00452	.00001				
2.6	.00570	.00001				
2.7	.00710	.00001				
2.8	.00876	.00002	.00001			
2.9	.01070	.00002	.00001			
3.0	.01295	.00003	.00001			
3.1	.01554	.00004	.00002	.00001		
3.2	.01849	.00006	.00002	.00001		
3.3	.02185	.00008	.00003	.00001		
3.4	.02564	.00010	.00004	.00002	.00001	
3.5	.02989	.00013	.00005	.00002	.00001	
3.6	.03462	.00018	.00007	.00003	.00001	.00001
3.7	.03988	.00023	.00010	.00004	.00002	.00001
3.8	.04568	.00030	.00013	.00006	.00002	.00001
3.9	.05206	.00038	.00017	.00007	.00003	.00001
4.0	.05904	.00049	.00022	.00010	.00004	.00002
4.1	.06666	.00062	.00028	.00013	.00006	.00003
4.2	.07492	.00078	.00037	.00017	.00008	.00004
4.3	.08387	.00099	.00047	.00023	.00011	.00005
4.4	.09352	.00124	.00061	.00029	.00014	.00007
4.5	.10389	.00156	.00077	.00038	.00019	.00009
4.6	.11500	.00194	.00099	.00050	.00025	.00013
4.7	.12688	.00242	.00125	.00065	.00033	.00017
4.8	.13954	.00300	.00158	.00083	.00044	.00023
4.9	.15300	.00371	.00199	.00107	.00058	.00031
5.0	.16727	.00457	.00251	.00138	.00076	.00041
5.1	.18236	.00562	.00315	.00176	.00099	.00055
5.2	.19829	.00689	.00393	.00225	.00128	.00073
5.3	.21507	.00842	.00491	.00286	.00167	.00097
5.4	.23271	.01028	.00611	.00363	.00216	.00128
5.5	.25122	.01251	.00759	.00460	.00279	.00169
5.6	.27060	.01519	.00940	.00582	.00360	.00223
5.7	.29086	.01841	.01162	.00734	.00463	.00292
5.8	.31201	.02227	.01434	.00924	.00595	.00383
5.9	.33405	.02688	.01766	.01160	.00762	.00501
6.0	.35698	.03238	.02171	.01455	.00975	.00654

Table 4-8. Erlang C Delay Loss Probability (*N* = 9 Servers)

Offered Traffic (*A* in Erl)	P (>0)	Delay Loss Probability *P* (>*t*) for T_1/T_2 =				
		0.2	0.4	0.6	0.8	1.0
3.1	.00501	.00154	.00047	.00015	.00004	.00001
3.2	.00613	.00192	.00060	.00019	.00006	.00002
3.3	.00744	.00238	.00076	.00024	.00008	.00002
3.4	.00897	.00293	.00095	.00031	.00010	.00003
3.5	.01072	.00357	.00119	.00040	.00013	.00004
3.6	.01273	.00432	.00147	.00050	.00017	.00006
3.7	.01502	.00520	.00180	.00062	.00022	.00007
3.8	.01760	.00622	.00220	.00078	.00027	.00010
3.9	.02051	.00739	.00267	.00096	.00035	.00013
4.0	.02376	.00874	.00322	.00118	.00044	.00016
4.1	.02738	.01028	.00386	.00145	.00054	.00020
4.2	.03140	.01202	.00460	.00176	.00067	.00026
4.3	.03583	.01400	.00547	.00214	.00083	.00033
4.4	.04071	.01622	.00647	.00258	.00103	.00041
4.5	.04605	.01872	.00761	.00309	.00126	.00051
4.6	.05188	.02152	.00893	.00370	.00154	.00064
4.7	.05822	.02464	.01042	.00441	.00187	.00079
4.8	.06509	.02810	.01213	.00524	.00226	.00098
4.9	.07251	.03194	.01407	.00620	.00273	.00120
5.0	.08051	.03625	.01625	.00730	.00328	.00147
5.1	.08910	.04084	.01872	.00858	.00393	.00180
5.2	.09831	.04597	.02150	.01006	.00470	.00220
5.3	.10814	.05160	.02462	.01175	.00560	.00267
5.4	.11862	.05774	.02811	.01368	.00666	.00324
5.5	.12977	.06444	.03200	.01589	.00789	.00392
5.6	.14160	.07174	.03634	.01841	.00933	.00473
5.7	.15412	.07966	.04117	.02128	.01100	.00458
5.8	.16735	.08824	.04653	.02453	.01294	.00682
5.9	.18130	.09753	.05246	.02822	.01518	.00817
6.0	.19598	.10756	.05903	.03240	.01778	.00976
6.1	.21140	.11837	.06627	.03711	.02078	.01163
6.2	.22758	.13000	.07426	.04242	.02423	.01384
6.3	.24452	.14249	.08304	.04839	.02820	.01643
6.4	.26223	.15590	.09269	.05510	.03276	.01948
6.5	.28071	.17026	.10327	.06263	.03799	.02304
6.6	.29997	.18562	.11486	.07107	.04398	.02721
6.7	.32003	.20203	.12754	.08051	.05083	.03209
6.8	.34087	.21953	.14139	.09106	.05864	.03777
6.9	.36251	.23819	.15650	.10283	.06756	.04439
7.0	.38495	.25804	.17297	.11594	.07772	.05210

(*table continues*)

Table 4-8. Erlang C Delay Loss Probability (N = 9 Servers *Continued*)

Offered Traffic (A in Erl)	P (>0)	Delay Loss Probability P (>t) for T_1/T_2 =				
		1.2	1.4	1.6	1.8	2.0
3.1	.00501					
3.2	.00613	.00001				
3.3	.00744	.00001				
3.4	.00897	.00001				
3.5	.01072	.00001				
3.6	.01273	.00002	.00001			
3.7	.01502	.00003	.00001			
3.8	.01760	.00003	.00001			
3.9	.02051	.00005	.00002	.00001		
4.0	.02376	.00006	.00002	.00001		
4.1	.02738	.00008	.00003	.00001		
4.2	.03140	.00010	.00004	.00001	.00001	
4.3	.03583	.00013	.00005	.00002	.00001	
4.4	.04071	.00016	.00006	.00003	.00001	
4.5	.04605	.00021	.00008	.00003	.00001	
4.6	.05188	.00026	.00011	.00005	.00002	.00001
4.7	.05822	.00033	.00014	.00006	.00003	.00001
4.8	.06509	.00042	.00018	.00008	.00003	.00001
4.9	.07251	.00053	.00023	.00010	.00005	.00002
5.0	.08051	.00066	.00030	.00013	.00006	.00003
5.1	.08910	.00083	.00038	.00017	.00008	.00004
5.2	.09831	.00103	.00048	.00022	.00011	.00005
5.3	.10814	.00128	.00061	.00029	.00014	.00007
5.4	.11862	.00158	.00077	.00037	.00018	.00009
5.5	.12977	.00195	.00097	.00048	.00024	.00012
5.6	.14160	.00239	.00121	.00061	.00031	.00016
5.7	.15412	.00294	.00152	.00078	.00041	.00021
5.8	.16735	.00360	.00190	.00100	.00053	.00028
5.9	.18130	.00439	.00236	.00127	.00068	.00037
6.0	.19598	.00535	.00294	.00161	.00089	.00049
6.1	.21140	.00651	.00365	.00204	.00114	.00064
6.2	.22758	.00791	.00452	.00258	.00147	.00084
6.3	.24452	.00958	.00558	.00325	.00190	.00110
6.4	.26223	.01158	.00688	.00409	.00243	.00145
6.5	.28071	.01398	.00848	.00514	.00312	.00189
6.6	.29997	.01684	.01042	.00645	.00399	.00247
6.7	.32003	.02025	.01279	.00807	.00510	.00322
6.8	.34087	.02432	.01567	.01009	.00650	.00418
6.9	.36251	.02917	.01916	.01259	.00827	.00544
7.0	.38495	.03492	.02341	.01569	.01052	.00705

Table 4-9. Erlang C Delay Loss Probability (N = 10 Servers)

Offered Traffic (A in Erl)	P (>0)	Delay Loss Probability P (>t) for T_1/T_2 =				
		0.2	0.4	0.6	0.8	1.0
4.1	.01038	.00319	.00098	.00030	.00009	.00003
4.2	.01216	.00381	.00119	.00037	.00012	.00004
4.3	.01416	.00453	.00145	.00046	.00015	.00005
4.4	.01641	.00535	.00175	.00057	.00019	.00006
4.5	.01892	.00630	.00210	.00070	.00023	.00008
4.6	.02171	.00737	.00250	.00085	.00029	.00010
4.7	.02481	.00859	.00298	.00103	.00036	.00012
4.8	.02823	.00998	.00353	.00125	.00044	.00016
4.9	.03199	.01153	.00416	.00150	.00054	.00020
5.0	.03611	.01328	.00489	.00180	.00066	.00024
5.1	.04061	.01524	.00572	.00215	.00081	.00030
5.2	.04550	.01742	.00667	.00255	.00098	.00037
5.3	.05082	.01985	.00775	.00303	.00118	.00046
5.4	.05658	.02255	.00899	.00358	.00143	.00057
5.5	.06279	.02553	.01038	.00422	.00172	.00070
5.6	.06947	.02882	.01195	.00496	.00206	.00085
5.7	.07665	.03244	.01373	.00581	.00246	.00104
5.8	.08433	.03641	.01572	.00679	.00293	.00126
5.9	.09255	.04076	.01795	.00791	.00348	.00153
6.0	.10130	.04552	.02045	.00919	.00413	.00186
6.1	.11061	.05070	.02324	.01065	.00488	.00224
6.2	.12049	.05635	.02635	.01232	.00576	.00270
6.3	.13096	.06248	.02981	.01422	.00679	.00324
6.4	.14203	.06913	.03365	.01638	.00797	.00388
6.5	.15371	.07633	.03791	.01882	.00935	.00464
6.6	.16602	.08411	.04261	.02159	.01094	.00554
6.7	.17897	.09250	.04781	.02471	.01277	.00660
6.8	.19256	.10154	.05354	.02823	.01489	.00787
6.9	.20681	.11125	.05985	.03219	.01732	.00932
7.0	.22173	.12169	.06678	.03665	.02011	.01104
7.1	.23732	.13288	.07440	.04166	.02332	.01306
7.2	.25360	.14486	.08275	.04726	.02700	.01542
7.3	.27057	.15768	.09189	.05355	.03120	.01818
7.4	.28824	.17136	.10188	.06057	.03601	.02141
7.5	.30661	.18597	.11280	.06841	.04150	.02517
7.6	.32569	.20153	.12470	.07716	.04775	.02955
7.7	.34548	.21810	.13768	.08692	.05487	.03464
7.8	.36599	.23571	.15181	.09777	.06297	.04055
7.9	.38722	.25442	.16717	.10984	.07217	.04742
8.0	.40918	.27428	.18386	.12324	.08261	.05538

(*table continues*)

Table 4-9. Erlang C Delay Loss Probability ($N = 10$ Servers *Continued*)

Offered Traffic (A in Erl)	$P(>0)$	Delay Loss Probability $P(>t)$ for $T_1/T_2 =$				
		1.2	1.4	1.6	1.8	2.0
4.1	.01038	.00001				
4.2	.01216	.00001				
4.3	.01416	.00002				
4.4	.01641	.00002	.00001			
4.5	.01892	.00003	.00001			
4.6	.02171	.00003	.00001			
4.7	.02481	.00004	.00001	.00001		
4.8	.02823	.00006	.00002	.00001		
4.9	.03199	.00007	.00003	.00001		
5.0	.03611	.00009	.00003	.00001		
5.1	.04061	.00011	.00004	.00002	.00001	
5.2	.04550	.00014	.00005	.00002	.00001	
5.3	.05082	.00018	.00007	.00003	.00001	
5.4	.05658	.00023	.00009	.00004	.00001	
5.5	.06279	.00028	.00012	.00005	.00002	.00001
5.6	.06947	.00035	.00015	.00006	.00003	.00001
5.7	.07665	.00044	.00019	.00008	.00003	.00001
5.8	.08433	.00055	.00024	.00010	.00004	.00002
5.9	.09255	.00068	.00030	.00013	.00006	.00003
6.0	.10130	.00083	.00037	.00017	.00008	.00003
6.1	.11061	.00103	.00047	.00022	.00010	.00005
6.2	.12049	.00126	.00059	.00028	.00013	.00006
6.3	.13096	.00154	.00074	.00035	.00017	.00008
6.4	.14203	.00189	.00092	.00045	.00022	.00011
6.5	.15371	.00231	.00114	.00057	.00028	.00014
6.6	.16602	.00281	.00142	.00072	.00036	.00018
6.7	.17897	.00341	.00176	.00091	.00047	.00024
6.8	.19256	.00414	.00218	.00115	.00061	.00032
6.9	.20681	.00501	.00270	.00145	.00078	.00042
7.0	.22173	.00606	.00332	.00182	.00100	.00055
7.1	.23732	.00731	.00409	.00229	.00128	.00072
7.2	.25360	.00881	.00503	.00287	.00164	.00094
7.3	.27057	.01060	.00618	.00360	.00210	.00122
7.4	.28824	.01273	.00757	.00450	.00267	.00159
7.5	.30661	.01527	.00926	.00562	.00341	.00207
7.6	.32569	.01828	.01131	.00700	.00433	.00268
7.7	.34548	.02187	.01380	.00871	.00550	.00347
7.8	.36599	.02612	.01682	.01083	.00698	.00449
7.9	.38722	.03116	.02047	.01345	.00884	.00581
8.0	.40918	.03712	.02488	.01668	.01118	.00749

Table 4-10. Erlang C Delay Loss Probability (*N* = 11 Servers)

Offered Traffic (*A* in Erl)	*P* (>0)	Delay Loss Probability *P* (>*t*) for T_1/T_2 =				
		0.2	0.4	0.6	0.8	1.0
5.1	.01726	.00530	.00163	.00050	.00015	.00005
5.2	.01966	.00616	.00193	.00061	.00019	.00006
5.3	.02232	.00714	.00228	.00073	.00023	.00007
5.4	.02523	.00823	.00269	.00088	.00029	.00009
5.5	.02843	.00946	.00315	.00105	.00035	.00012
5.6	.03193	.01084	.00368	.00125	.00042	.00014
5.7	.03574	.01238	.00429	.00149	.00051	.00018
5.8	.03988	.01410	.00498	.00176	.00062	.00022
5.9	.04437	.01600	.00577	.00208	.00075	.00027
6.0	.04922	.01811	.00666	.00245	.00090	.00033
6.1	.05445	.02044	.00767	.00288	.00108	.00041
6.2	.06008	.02300	.00881	.00337	.00129	.00049
6.3	.06611	.02583	.01009	.00394	.00154	.00060
6.4	.07257	.02892	.01153	.00459	.00183	.00073
6.5	.07948	.03231	.01314	.00534	.00217	.00088
6.6	.08683	.03602	.01494	.00620	.00257	.00107
6.7	.09466	.04006	.01695	.00717	.00304	.00128
6.8	.10298	.04446	.01919	.00829	.00358	.00154
6.9	.11179	.04923	.02168	.00955	.00421	.00185
7.0	.12111	.05442	.02445	.01099	.00494	.00222
7.1	.13095	.06003	.02752	.01261	.00578	.00265
7.2	.14133	.06610	.03091	.01446	.00676	.00316
7.3	.15226	.07265	.03466	.01654	.00789	.00376
7.4	.16375	.07971	.03880	.01888	.00919	.00447
7.5	.17581	.08730	.04335	.02153	.01069	.00531
7.6	.18844	.09547	.04837	.02450	.01241	.00629
7.7	.20167	.10423	.05387	.02784	.01439	.00744
7.8	.21549	.11363	.05991	.03159	.01666	.00878
7.9	.22992	.12368	.06653	.03579	.01923	.01036
8.0	.24496	.13444	.07378	.04049	.02222	.01220
8.1	.26062	.14592	.08170	.04574	.02561	.01434
8.2	.27691	.15817	.09035	.05161	.02948	.01684
8.3	.29384	.17123	.09979	.05815	.03389	.01975
8.4	.31140	.18513	.11007	.06544	.03890	.02313
8.5	.32961	.19992	.12126	.07355	.04461	.02706
8.6	.34847	.21563	.13343	.08256	.05109	.03161
8.7	.36798	.23230	.14665	.09258	.05844	.03689
8.8	.38815	.24998	.16100	.10369	.06678	.04301
8.9	.40898	.26872	.17656	.11601	.07622	.05008
9.0	.43047	.28855	.19342	.12966	.08691	.05826

(table continues)

Table 4-10. Erlang C Delay Loss Probability ($N = 11$ Servers *Continued*)

Offered Traffic (*A* in Erl)	P (>0)	Delay Loss Probability P (>t) for $T_1/T_2 =$				
		1.2	1.4	1.6	1.8	2.0
5.1	.01726	.00001				
5.2	.01966	.00002	.00001			
5.3	.02232	.00002	.00001			
5.4	.02523	.00003	.00001			
5.5	.02843	.00004	.00001			
5.6	.03193	.00005	.00002	.00001		
5.7	.03574	.00006	.00002	.00001		
5.8	.03988	.00008	.00003	.00001		
5.9	.04437	.00010	.00004	.00001		
6.0	.04922	.00012	.00004	.00002	.00001	
6.1	.05445	.00015	.00006	.00002	.00001	
6.2	.06008	.00019	.00007	.00003	.00001	
6.3	.06611	.00023	.00009	.00004	.00001	.00001
6.4	.07257	.00029	.00012	.00005	.00002	.00001
6.5	.07948	.00036	.00015	.00006	.00002	.00001
6.6	.08683	.00044	.00018	.00008	.00003	.00001
6.7	.09466	.00054	.00023	.00010	.00004	.00002
6.8	.10298	.00067	.00029	.00012	.00005	.00002
6.9	.11179	.00082	.00036	.00016	.00007	.00003
7.0	.12111	.00100	.00045	.00020	.00009	.00004
7.1	.13095	.00122	.00056	.00026	.00012	.00005
7.2	.14133	.00148	.00069	.00032	.00015	.00007
7.3	.15226	.00180	.00086	.00041	.00020	.00009
7.4	.16375	.00218	.00106	.00052	.00025	.00012
7.5	.17581	.00264	.00131	.00065	.00032	.00016
7.6	.18844	.00319	.00161	.00082	.00041	.00021
7.7	.20167	.00384	.00199	.00103	.00053	.00027
7.8	.21549	.00463	.00244	.00129	.00068	.00036
7.9	.22992	.00557	.00300	.00161	.00087	.00047
8.0	.24496	.00669	.00367	.00202	.00111	.00061
8.1	.26062	.00803	.00450	.00252	.00141	.00079
8.2	.27691	.00962	.00549	.00314	.00179	.00102
8.3	.29384	.01151	.00671	.00391	.00228	.00133
8.4	.31140	.01375	.00817	.00486	.00289	.00172
8.5	.32961	.01641	.00995	.00604	.00366	.00222
8.6	.34847	.01956	.01210	.00749	.00463	.00287
8.7	.36798	.02329	.01470	.00928	.00586	.00370
8.8	.38815	.02770	.01784	.01149	.00740	.00477
8.9	.40898	.03291	.02162	.01421	.00933	.00613
9.0	.43047	.03905	.02618	.01755	.01176	.00788

Table 4-11. Erlang C Delay Loss Probability (N = 12 Servers)

Offered Traffic (A in Erl)	P (>0)	Delay Loss Probability P (>t) for T_1/T_2 =				
		0.2	0.4	0.6	0.8	1.0
6.1	.02521	.00775	.00238	.00073	.00022	.00007
6.2	.02819	.00884	.00277	.00087	.00027	.00009
6.3	.03143	.01005	.00321	.00103	.00033	.00011
6.4	.03495	.01140	.00372	.00121	.00040	.00013
6.5	.03876	.01290	.00429	.00143	.00048	.00016
6.6	.04287	.01456	.00494	.00168	.00057	.00019
6.7	.04729	.01639	.00568	.00197	.00068	.00024
6.8	.05205	.01840	.00650	.00230	.00081	.00029
6.9	.05716	.02061	.00743	.00268	.00097	.00035
7.0	.06262	.02304	.00847	.00312	.00115	.00042
7.1	.06846	.02569	.00964	.00362	.00136	.00051
7.2	.07468	.02859	.01095	.00419	.00161	.00061
7.3	.08130	.03176	.01240	.00485	.00189	.00074
7.4	.08833	.03520	.01403	.00559	.00223	.00089
7.5	.09578	.03894	.01583	.00644	.00262	.00106
7.6	.10367	.04300	.01784	.00740	.00307	.00127
7.7	.11201	.04740	.02006	.00849	.00359	.00152
7.8	.12081	.05216	.02252	.00972	.00420	.00181
7.9	.13009	.05729	.02523	.01111	.00489	.00216
8.0	.13984	.06283	.02823	.01269	.00570	.00256
8.1	.15009	.06880	.03154	.01446	.00663	.00304
8.2	.16084	.07522	.03518	.01645	.00769	.00360
8.3	.17210	.08211	.03918	.01869	.00892	.00425
8.4	.18389	.08951	.04357	.02121	.01032	.00502
8.5	.19621	.09743	.04838	.02403	.01193	.00592
8.6	.20906	.10591	.05366	.02718	.01377	.00698
8.7	.22247	.11498	.05943	.03072	.01588	.00821
8.8	.23642	.12466	.06573	.03466	.01828	.00964
8.9	.25094	.13499	.07262	.03907	.02101	.01130
9.0	.26603	.14600	.08013	.04398	.02413	.01325
9.1	.28170	.15772	.08831	.04944	.02768	.01550
9.2	.29795	.17019	.09721	.05553	.03172	.01812
9.3	.31478	.18344	.10690	.06229	.03630	.02115
9.4	.33220	.19750	.11742	.06981	.04150	.02467
9.5	.35022	.21242	.12884	.07815	.04740	.02875
9.6	.36884	.22823	.14123	.08739	.05407	.03346
9.7	.38807	.24498	.15465	.09763	.06163	.03891
9.8	.40790	.26270	.16919	.10896	.07018	.04520
9.9	.42834	.28144	.18492	.12150	.07983	.05245
10.0	.44939	.30123	.20192	.13535	.09073	.06082

(table continues)

Table 4-11. Erlang C Delay Loss Probability (N = 12 Servers *Continued*)

Offered Traffic (A in Erl)	P (>0)	Delay Loss Probability P (>t) for T_1/T_2 =				
		1.2	1.4	1.6	1.8	2.0
6.1	.02521	.00002	.00001			
6.2	.02819	.00003	.00001			
6.3	.03143	.00003	.00001			
6.4	.03495	.00004	.00001			
6.5	.03876	.00005	.00002	.00001		
6.6	.04287	.00007	.00002	.00001		
6.7	.04729	.00008	.00003	.00001		
6.8	.05205	.00010	.00004	.00001		
6.9	.05716	.00013	.00005	.00002	.00001	
7.0	.06262	.00016	.00006	.00002	.00001	
7.1	.06846	.00019	.00007	.00003	.00001	
7.2	.07468	.00024	.00009	.00003	.00001	
7.3	.08130	.00029	.00011	.00004	.00002	.00001
7.4	.08833	.00035	.00014	.00006	.00002	.00001
7.5	.09578	.00043	.00018	.00007	.00003	.00001
7.6	.10367	.00053	.00022	.00009	.00004	.00001
7.7	.11201	.00064	.00027	.00012	.00005	.00002
7.8	.12081	.00078	.00034	.00015	.00006	.00003
7.9	.13009	.00095	.00042	.00018	.00008	.00004
8.0	.13984	.00115	.00052	.00023	.00010	.00005
8.1	.15009	.00139	.00064	.00029	.00013	.00006
8.2	.16084	.00168	.00079	.00037	.00017	.00008
8.3	.17210	.00203	.00097	.00046	.00022	.00011
8.4	.18389	.00245	.00119	.00058	.00028	.00014
8.5	.19621	.00294	.00146	.00073	.00036	.00018
8.6	.20906	.00353	.00179	.00091	.00046	.00023
8.7	.22247	.00424	.00219	.00113	.00059	.00030
8.8	.23642	.00508	.00268	.00141	.00074	.00039
8.9	.25094	.00608	.00327	.00176	.00095	.00051
9.0	.26603	.00727	.00399	.00219	.00120	.00066
9.1	.28170	.00868	.00486	.00272	.00152	.00085
9.2	.29795	.01035	.00591	.00338	.00193	.00110
9.3	.31478	.01233	.00718	.00419	.00244	.00142
9.4	.33220	.01467	.00872	.00518	.00308	.00183
9.5	.35022	.01744	.01058	.00641	.00389	.00236
9.6	.36884	.02070	.01281	.00793	.00491	.00304
9.7	.38807	.02456	.01551	.00979	.00618	.00390
9.8	.40790	.02911	.01875	.01207	.00778	.00501
9.9	.42834	.03446	.02264	.01488	.00978	.00642
10.0	.44939	.04077	.02733	.01832	.01228	.00823

Table 4-12. Erlang C Delay Loss Probability (N = 13 Servers)

Offered Traffic (A in Erl)	P (>0)	Delay Loss Probability P (>t) for T_1/T_2 =				
		0.2	0.4	0.6	0.8	1.0
7.1	.03387	.01041	.00320	.00098	.00030	.00009
7.2	.03737	.01171	.00367	.00115	.00036	.00011
7.3	.04114	.01316	.00421	.00135	.00043	.00014
7.4	.04518	.01474	.00481	.00157	.00051	.00017
7.5	.04952	.01648	.00549	.00183	.00061	.00020
7.6	.05416	.01839	.00625	.00212	.00072	.00024
7.7	.05912	.02048	.00710	.00246	.00085	.00030
7.8	.06440	.02276	.00805	.00284	.00101	.00036
7.9	.07002	.02525	.00911	.00328	.00118	.00043
8.0	.07600	.02796	.01029	.00378	.00139	.00051
8.1	.08234	.03090	.01160	.00435	.00163	.00061
8.2	.08905	.03410	.01306	.00500	.00191	.00073
8.3	.09615	.03756	.01467	.00573	.00224	.00087
8.4	.10364	.04130	.01646	.00656	.00261	.00104
8.5	.11154	.04535	.01844	.00750	.00305	.00124
8.6	.11986	.04971	.02062	.00855	.00355	.00147
8.7	.12860	.05442	.02303	.00974	.00412	.00174
8.8	.13779	.05948	.02568	.01109	.00479	.00207
8.9	.14741	.06493	.02860	.01259	.00555	.00244
9.0	.15750	.07077	.03180	.01429	.00642	.00288
9.1	.16805	.07704	.03531	.01619	.00742	.00340
9.2	.17908	.08375	.03917	.01832	.00857	.00401
9.3	.19059	.09093	.04338	.02070	.00988	.00471
9.4	.20258	.09861	.04800	.02336	.01137	.00554
9.5	.21508	.10681	.05304	.02634	.01308	.00649
9.6	.22808	.11555	.05854	.02966	.01502	.00761
9.7	.24160	.12487	.06454	.03336	.01728	.00891
9.8	.25563	.13479	.07107	.03748	.01976	.01042
9.9	.27018	.14534	.07819	.04206	.02263	.01217
10.0	.28527	.15656	.08592	.04716	.02588	.01420
10.1	.30089	.16847	.09433	.05281	.02957	.01656
10.2	.31706	.18111	.10345	.05909	.03375	.01928
10.3	.33377	.19450	.11335	.06605	.03849	.02243
10.4	.35103	.20870	.12407	.07376	.04385	.02607
10.5	.36885	.22372	.13569	.08230	.04992	.03028
10.6	.38722	.23961	.14826	.09174	.05677	.03513
10.7	.40616	.25640	.16186	.10218	.06451	.04072
10.8	.42566	.27414	.17656	.11371	.07323	.04716
10.9	.44572	.29286	.19242	.12643	.08307	.05458
11.0	.46636	.31261	.20955	.14047	.09416	.06312

(*table continues*)

Table 4-12. Erlang C Delay Loss Probability ($N = 13$ Servers *Continued*)

Offered Traffic (A in Erl)	P (>0)	Delay Loss Probability P (>t) for T_1/T_2 =				
		1.2	1.4	1.6	1.8	2.0
7.1	.03387	.00003	.00001			
7.2	.03737	.00004	.00001			
7.3	.04114	.00004	.00001			
7.4	.04518	.00005	.00002	.00001		
7.5	.04952	.00007	.00002	.00001		
7.6	.05416	.00008	.00003	.00001		
7.7	.05912	.00010	.00004	.00001		
7.8	.06440	.00013	.00004	.00002	.00001	
7.9	.07002	.00015	.00006	.00002	.00001	
8.0	.07600	.00019	.00007	.00003	.00001	
8.1	.08234	.00023	.00009	.00003	.00001	
8.2	.08905	.00028	.00011	.00004	.00002	.00001
8.3	.09615	.00034	.00013	.00005	.00002	.00001
8.4	.10364	.00042	.00017	.00007	.00003	.00001
8.5	.11154	.00050	.00020	.00008	.00003	.00001
8.6	.11986	.00061	.00025	.00011	.00004	.00002
8.7	.12860	.00074	.00031	.00013	.00006	.00002
8.8	.13779	.00089	.00039	.00017	.00007	.00003
8.9	.14741	.00108	.00047	.00021	.00009	.00004
9.0	.15750	.00130	.00058	.00026	.00012	.00005
9.1	.16805	.00156	.00071	.00033	.00015	.00007
9.2	.17908	.00187	.00088	.00041	.00019	.00009
9.3	.19059	.00225	.00107	.00051	.00024	.00012
9.4	.20258	.00269	.00131	.00064	.00031	.00015
9.5	.21508	.00323	.00160	.00080	.00039	.00020
9.6	.22808	.00386	.00195	.00099	.00050	.00025
9.7	.24160	.00461	.00238	.00123	.00064	.00033
9.8	.25563	.00549	.00290	.00153	.00081	.00042
9.9	.27018	.00655	.00352	.00189	.00102	.00055
10.0	.28527	.00779	.00428	.00235	.00129	.00071
10.1	.30089	.00927	.00519	.00291	.00163	.00091
10.2	.31706	.01101	.00629	.00359	.00205	.00117
10.3	.33377	.01307	.00762	.00444	.00259	.00151
10.4	.35103	.01550	.00922	.00548	.00326	.00194
10.5	.36885	.01836	.01114	.00676	.00410	.00249
10.6	.38722	.02174	.01345	.00832	.00515	.00319
10.7	.40616	.02571	.01623	.01024	.00647	.00408
10.8	.42566	.03038	.01956	.01260	.00811	.00523
10.9	.44572	.03586	.02356	.01548	.01017	.00668
11.0	.46636	.04231	.02836	.01901	.01274	.00854

Table 4-13. Erlang C Delay Loss Probability (N = 14 Servers)

Offered Traffic (A in Erl)	P (>0)	Delay Loss Probability P (>t) for T_1/T_2 =				
		0.2	0.4	0.6	0.8	1.0
8.1	.04298	.01321	.00406	.00125	.00038	.00012
8.2	.04694	.01471	.00461	.00145	.00045	.00014
8.3	.05117	.01636	.00523	.00167	.00054	.00017
8.4	.05567	.01817	.00593	.00193	.00063	.00021
8.5	.06047	.02013	.00670	.00223	.00074	.00025
8.6	.06557	.02227	.00756	.00257	.00087	.00030
8.7	.07098	.02459	.00852	.00295	.00102	.00035
8.8	.07671	.02711	.00958	.00339	.00120	.00042
8.9	.08277	.02985	.01076	.00388	.00140	.00050
9.0	.08918	.03281	.01207	.00444	.00163	.00060
9.1	.09593	.03600	.01351	.00507	.00190	.00071
9.2	.10305	.03946	.01511	.00578	.00221	.00085
9.3	.11054	.04318	.01687	.00659	.00257	.00101
9.4	.11841	.04719	.01881	.00749	.00299	.00119
9.5	.12667	.05150	.02094	.00851	.00346	.00141
9.6	.13533	.05613	.02328	.00966	.00401	.00166
9.7	.14440	.06110	.02586	.01094	.00463	.00196
9.8	.15388	.06643	.02868	.01238	.00535	.00231
9.9	.16379	.07214	.03177	.01399	.00616	.00271
10.0	.17413	.07824	.03516	.01580	.00710	.00319
10.1	.18491	.08477	.03886	.01781	.00817	.00374
10.2	.19614	.09173	.04290	.02006	.00938	.00439
10.3	.20783	.09916	.04731	.02257	.01077	.00514
10.4	.21998	.10708	.05212	.02537	.01235	.00601
10.5	.23259	.11550	.05736	.02848	.01414	.00702
10.6	.24569	.12447	.06306	.03195	.01618	.00820
10.7	.25926	.13400	.06926	.03580	.01850	.00956
10.8	.27332	.14412	.07599	.04007	.02113	.01114
10.9	.28787	.15486	.08331	.04481	.02411	.01297
11.0	.30292	.16625	.09124	.05007	.02748	.01508
11.1	.31847	.17831	.09984	.05590	.03130	.01752
11.2	.33453	.19109	.10915	.06235	.03561	.02034
11.3	.35110	.20460	.11923	.06948	.04049	.02360
11.4	.36819	.21890	.13014	.07737	.04600	.02735
11.5	.38579	.23400	.14193	.08608	.05221	.03167
11.6	.40392	.24994	.15466	.09570	.05922	.03664
11.7	.42257	.26679	.16840	.10631	.06711	.04237
11.8	.44175	.28451	.18323	.11801	.07600	.04895
11.9	.46146	.30320	.19922	.13090	.08601	.05651
12.0	.48171	.32290	.21645	.14509	.09726	.06519

(*table continues*)

Table 4-13. Erlang C Delay Loss Probability (N = 14 Servers *Continued*)

Offered Traffic (A in Erl)	P (>0)	Delay Loss Probability P (>t) for T_1/T_2 =				
		1.2	1.4	1.6	1.8	2.0
8.1	.04298	.00004	.00001			
8.2	.04694	.00004	.00001			
8.3	.05117	.00005	.00002	.00001		
8.4	.05567	.00007	.00002	.00001		
8.5	.06047	.00008	.00003	.00001		
8.6	.06557	.00010	.00003	.00001		
8.7	.07098	.00012	.00004	.00001	.00001	
8.8	.07671	.00015	.00005	.00002	.00001	
8.9	.08277	.00018	.00007	.00002	.00001	
9.0	.08918	.00022	.00008	.00003	.00001	
9.1	.09593	.00027	.00010	.00004	.00001	.00001
9.2	.10305	.00032	.00012	.00005	.00002	.00001
9.3	.11054	.00039	.00015	.00006	.00002	.00001
9.4	.11841	.00047	.00019	.00008	.00003	.00001
9.5	.12667	.00057	.00023	.00009	.00004	.00002
9.6	.13533	.00069	.00029	.00012	.00005	.00002
9.7	.14440	.00083	.00035	.00015	.00006	.00003
9.8	.15388	.00100	.00043	.00019	.00008	.00003
9.9	.16379	.00120	.00053	.00023	.00010	.00004
10.0	.17413	.00143	.00064	.00029	.00013	.00006
10.1	.18491	.00172	.00079	.00036	.00017	.00008
10.2	.19614	.00205	.00096	.00045	.00021	.00010
10.3	.20783	.00245	.00117	.00056	.00027	.00013
10.4	.21998	.00293	.00142	.00069	.00034	.00016
10.5	.23259	.00349	.00173	.00086	.00043	.00021
10.6	.24569	.00415	.00210	.00107	.00054	.00027
10.7	.25926	.00494	.00255	.00132	.00068	.00035
10.8	.27332	.00587	.00310	.00163	.00086	.00045
10.9	.28787	.00698	.00375	.00202	.00109	.00058
11.0	.30292	.00828	.00454	.00249	.00137	.00075
11.1	.31847	.00981	.00549	.00308	.00172	.00096
11.2	.33453	.01162	.00664	.00379	.00217	.00124
11.3	.35110	.01375	.00801	.00467	.00272	.00159
11.4	.36819	.01626	.00967	.00575	.00342	.00203
11.5	.38579	.01921	.01165	.00707	.00429	.00260
11.6	.40392	.02267	.01403	.00868	.00537	.00332
11.7	.42257	.02675	.01688	.01066	.00673	.00425
11.8	.44175	.03152	.02030	.01308	.00842	.00542
11.9	.46146	.03713	.02440	.01603	.01053	.00692
12.0	.48171	.04370	.02929	.01964	.01316	.00882

Table 4-14. Erlang C Delay Loss Probability (N = 15 Servers)

Offered Traffic (A in Erl)	P (>0)	Delay Loss Probability P (>t) for T_1/T_2 =				
		0.2	0.4	0.6	0.8	1.0
9.1	.05234	.01608	.00494	.00152	.00047	.00014
9.2	.05671	.01778	.00557	.00175	.00055	.00017
9.3	.06134	.01962	.00627	.00201	.00064	.00021
9.4	.06625	.02162	.00705	.00230	.00075	.00024
9.5	.07144	.02378	.00792	.00264	.00088	.00029
9.6	.07694	.02613	.00887	.00301	.00102	.00035
9.7	.08273	.02866	.00993	.00344	.00119	.00041
9.8	.08884	.03140	.01110	.00392	.00139	.00049
9.9	.09528	.03436	.01239	.00447	.00161	.00058
10.0	.10204	.03754	.01381	.00508	.00187	.00069
10.1	.10915	.04097	.01537	.00577	.00217	.00081
10.2	.11661	.04465	.01710	.00655	.00251	.00096
10.3	.12442	.04860	.01899	.00742	.00290	.00113
10.4	.13260	.05284	.02106	.00839	.00334	.00133
10.5	.14115	.05739	.02333	.00949	.00386	.00157
10.6	.15009	.06225	.02582	.01071	.00444	.00184
10.7	.15941	.06746	.02885	.01208	.00511	.00216
10.8	.16914	.07302	.03152	.01361	.00588	.00254
10.9	.17926	.07895	.03477	.01532	.00675	.00297
11.0	.18980	.08528	.03832	.01722	.00774	.00348
11.1	.20076	.09203	.04219	.01934	.00886	.00406
11.2	.21214	.09921	.04640	.02170	.01015	.00475
11.3	.22395	.10685	.05098	.02432	.01161	.00554
11.4	.23620	.11497	.05596	.02724	.01326	.00645
11.5	.24889	.12360	.06138	.03048	.01514	.00752
11.6	.26204	.13275	.06725	.03407	.01726	.00875
11.7	.27563	.14246	.07363	.03806	.01967	.01017
11.8	.28969	.15275	.08054	.04247	.02239	.01181
11.9	.30421	.16365	.08803	.04736	.02548	.01370
12.0	.31919	.17518	.09614	.05276	.02896	.01589
12.1	.33465	.18737	.10491	.05874	.03289	.01841
12.2	.35059	.20026	.11439	.06534	.03732	.02132
12.3	.36701	.21388	.12464	.07263	.04233	.02467
12.4	.38391	.22825	.13570	.08067	.04796	.02852
12.5	.40131	.24340	.14763	.08954	.05431	.03294
12.6	.41919	.25939	.16051	.09932	.06146	.03803
12.7	.43756	.27623	.17438	.11008	.06949	.04387
12.8	.45644	.29396	.18932	.12193	.07853	.05058
12.9	.47581	.31263	.20541	.13497	.08868	.05827
13.0	.49568	.33227	.22273	.14930	.10008	.06708

(*table continues*)

Table 4-14. Erlang C Delay Loss Probability (N = 15 Servers *Continued*)

Offered Traffic (A in Erl)	P (>0)	Delay Loss Probability P (>t) for T_1/T_2 =				
		1.2	1.4	1.6	1.8	2.0
9.1	.05234	.00004	.00001			
9.2	.05671	.00005	.00002	.00001		
9.3	.06134	.00007	.00002	.00001		
9.4	.06625	.00008	.00003	.00001		
9.5	.07144	.00010	.00003	.00001		
9.6	.07694	.00012	.00004	.00001		
9.7	.08273	.00014	.00005	.00002	.00001	
9.8	.08884	.00017	.00006	.00002	.00001	
9.9	.09528	.00028	.00008	.00003	.00001	
10.0	.10204	.00025	.00009	.00003	.00001	
10.1	.10915	.00031	.00011	.00004	.00002	.00001
10.2	.11661	.00037	.00014	.00005	.00002	.00001
10.3	.12442	.00044	.00017	.00007	.00003	.00001
10.4	.13260	.00053	.00021	.00008	.00003	.00001
10.5	.14115	.00064	.00026	.00011	.00004	.00002
10.6	.15009	.00076	.00032	.00013	.00005	.00002
10.7	.15941	.00092	.00039	.00016	.00007	.00003
10.8	.16914	.00109	.00047	.00020	.00009	.00004
10.9	.17926	.00131	.00058	.00025	.00011	.00005
11.0	.18980	.00156	.00070	.00032	.00014	.00006
11.1	.20076	.00186	.00085	.00039	.00018	.00008
11.2	.21214	.00222	.00104	.00049	.00023	.00011
11.3	.22395	.00264	.00126	.00060	.00029	.00014
11.4	.23620	.00314	.00153	.00074	.00036	.00018
11.5	.24889	.00373	.00185	.00092	.00046	.00023
11.6	.26204	.00443	.00224	.00114	.00058	.00029
11.7	.27563	.00525	.00272	.00140	.00073	.00037
11.8	.28969	.00623	.00328	.00173	.00091	.00048
11.9	.30421	.00737	.00397	.00213	.00115	.00062
12.0	.31919	.00872	.00479	.00263	.00144	.00079
12.1	.33465	.01031	.00577	.00323	.00181	.00101
12.2	.35059	.01218	.00696	.00397	.00227	.00130
12.3	.36701	.01437	.00838	.00488	.00284	.00166
12.4	.38391	.01695	.01008	.00599	.00356	.00212
12.5	.40131	.01998	.01212	.00735	.00446	.00270
12.6	.41919	.02353	.01456	.00901	.00558	.00345
12.7	.43756	.02769	.01748	.01104	.00697	.00440
12.8	.45644	.03257	.02098	.01351	.00870	.00560
12.9	.47581	.03828	.02515	.01653	.01086	.00714
13.0	.49568	.04497	.03014	.02021	.01354	.00908

Table 4-15. Erlang C Delay Loss Probability (N = 16 Servers)

Offered Traffic (A in Erl)	P (>0)	Delay Loss Probability P (>t) for T_1/T_2 =				
		0.2	0.4	0.6	0.8	1.0
10.1	.06181	.01899	.00584	.00179	.00055	.00017
10.2	.06653	.02086	.00654	.00205	.00064	.00020
10.3	.07152	.02287	.00732	.00234	.00075	.00024
10.4	.07678	.02505	.00817	.00267	.00087	.00028
10.5	.08232	.02740	.00912	.00304	.00101	.00034
10.6	.08815	.02994	.01017	.00345	.00117	.00040
10.7	.09428	.03266	.01132	.00392	.00136	.00047
10.8	.10072	.03560	.01258	.00445	.00157	.00056
10.9	.10747	.03875	.01397	.00504	.00182	.00066
11.0	.11454	.04214	.01550	.00570	.00210	.00077
11.1	.12194	.04576	.01718	.00645	.00242	.00091
11.2	.12968	.04965	.01901	.00728	.00279	.00107
11.3	.13776	.05381	.02102	.00821	.00321	.00125
11.4	.14619	.05826	.02322	.00925	.00369	.00147
11.5	.15499	.06301	.02562	.01042	.00423	.00172
11.6	.16415	.06808	.02824	.01171	.00486	.00202
11.7	.17368	.07349	.03110	.01316	.00557	.00236
11.8	.18359	.07926	.03422	.01477	.00638	.00275
11.9	.19389	.08539	.03761	.01656	.00730	.00321
12.0	.20458	.09192	.04130	.01856	.00834	.00375
12.1	.21566	.09886	.04532	.02077	.00952	.00437
12.2	.22715	.10623	.04968	.02323	.01087	.00508
12.3	.23905	.11406	.05442	.02596	.01239	.00591
12.4	.25137	.12236	.05956	.02899	.01411	.00678
12.5	.26411	.13115	.06513	.03234	.01606	.00798
12.6	.27727	.14047	.07116	.03605	.01827	.00925
12.7	.29086	.15033	.07770	.04016	.02076	.01073
12.8	.30489	.16076	.08477	.04470	.02357	.01243
12.9	.31935	.17179	.09242	.04971	.02674	.01439
13.0	.33426	.18345	.10068	.05525	.03032	.01664
13.1	.34962	.19575	.10960	.06137	.03436	.01924
13.2	.36542	.20873	.11923	.06811	.03890	.02222
13.3	.38168	.22243	.12962	.07554	.04402	.02565
13.4	.39840	.23686	.14082	.08372	.04977	.02959
13.5	.41558	.25206	.15288	.09273	.05624	.03411
13.6	.43322	.26807	.16588	.10264	.06351	.03930
13.7	.45133	.28492	.17987	.11355	.07168	.04525
13.8	.46991	.30264	.19491	.12553	.08085	.05207
13.9	.48896	.32127	.21109	.13870	.09113	.05988
14.0	.50849	.34085	.22848	.15315	.10266	.06882

(table continues)

Table 4-15. Erlang C Delay Loss Probability ($N = 16$ Servers *Continued*)

Offered Traffic (A in Erl)	$P(>0)$	Delay Loss Probability $P(>t)$ for $T_1/T_2 =$				
		1.2	1.4	1.6	1.8	2.0
10.1	.06181	.00005	.00002			
10.2	.06653	.00006	.00002	.00001		
10.3	.07152	.00008	.00002	.00001		
10.4	.07678	.00009	.00003	.00001		
10.5	.08232	.00011	.00004	.00001		
10.6	.08815	.00014	.00005	.00002	.00001	
10.7	.09428	.00016	.00006	.00002	.00001	
10.8	.10072	.00020	.00007	.00002	.00001	
10.9	.10747	.00024	.00009	.00003	.00001	
11.0	.11454	.00028	.00010	.00004	.00001	.00001
11.1	.12194	.00034	.00013	.00005	.00002	.00001
11.2	.12968	.00041	.00016	.00006	.00002	.00001
11.3	.13776	.00049	.00019	.00007	.00003	.00001
11.4	.14619	.00059	.00023	.00009	.00004	.00001
11.5	.15499	.00070	.00028	.00012	.00005	.00002
11.6	.16415	.00084	.00035	.00014	.00006	.00002
11.7	.17368	.00100	.00042	.00018	.00008	.00003
11.8	.18359	.00119	.00051	.00022	.00010	.00004
11.9	.19389	.00142	.00062	.00027	.00012	.00005
12.0	.20458	.00168	.00076	.00034	.00015	.00007
12.1	.21566	.00200	.00092	.00042	.00019	.00009
12.2	.22715	.00238	.00111	.00052	.00024	.00011
12.3	.23905	.00282	.00135	.00064	.00031	.00015
12.4	.25137	.00334	.00163	.00079	.00039	.00019
12.5	.26411	.00396	.00197	.00098	.00048	.00024
12.6	.27727	.00469	.00238	.00120	.00061	.00031
12.7	.29086	.00554	.00287	.00148	.00077	.00044
12.8	.30489	.00655	.00346	.00182	.00096	.00051
12.9	.31935	.00774	.00416	.00224	.00120	.00065
13.0	.33426	.00913	.00501	.00275	.00151	.00083
13.1	.34962	.01077	.00603	.00338	.00189	.00106
13.2	.36542	.01269	.00725	.00414	.00237	.00135
13.3	.38168	.01495	.00871	.00508	.00296	.00172
13.4	.39840	.01759	.01046	.00622	.00370	.00220
13.5	.41558	.02069	.01255	.00761	.00462	.00280
13.6	.43322	.02432	.01505	.00931	.00576	.00357
13.7	.45133	.02857	.01803	.01138	.00719	.00454
13.8	.46991	.03353	.02160	.01391	.00896	.00577
13.9	.48896	.03934	.02585	.01698	.01116	.00733
14.0	.50849	.04613	.03092	.02073	.01389	.00931

Table 4-16. Erlang C Delay Loss Probability (*N* = 17 Servers)

Offered Traffic (*A* in Erl)	*P* (>0)	Delay Loss Probability *P* (>*t*) for T_1/T_2 =				
		0.2	0.4	0.6	0.8	1.0
11.1	.07128	.02190	.00673	.00207	.00064	.00020
11.2	.07632	.02392	.00750	.00235	.00074	.00023
11.3	.08162	.02610	.00835	.00267	.00085	.00027
11.4	.08718	.02845	.00928	.00303	.00099	.00032
11.5	.09303	.03097	.01031	.00343	.00114	.00038
11.6	.09915	.03367	.01143	.00388	.00132	.00045
11.7	.10557	.03657	.01267	.00439	.00152	.00053
11.8	.11228	.03969	.01403	.00496	.00175	.00062
11.9	.11930	.04302	.01551	.00559	.00202	.00073
12.0	.12663	.04658	.01714	.00630	.00232	.00085
12.1	.13428	.05040	.01891	.00710	.00266	.00100
12.2	.14225	.05447	.02086	.00799	.00306	.00117
12.3	.15056	.05881	.02297	.00897	.00351	.00137
12.4	.15920	.06345	.02528	.01008	.00402	.00160
12.5	.16819	.06838	.02780	.01130	.00460	.00187
12.6	.17753	.07364	.03054	.01267	.00525	.00218
12.7	.18723	.07923	.03353	.01419	.00600	.00254
12.8	.19729	.08517	.03677	.01587	.00685	.00296
12.9	.20772	.09149	.04029	.01775	.00782	.00344
13.0	.21853	.09819	.04412	.01982	.00891	.00400
13.1	.22971	.10530	.04827	.02213	.01014	.00465
13.2	.24128	.11284	.05277	.02468	.01154	.00540
13.3	.25324	.12082	.05765	.02750	.01312	.00626
13.4	.26559	.12928	.06293	.03063	.01491	.00726
13.5	.27835	.13822	.06864	.03409	.01693	.00841
13.6	.29151	.14768	.07482	.03790	.01920	.00973
13.7	.30507	.15768	.08150	.04212	.02177	.01125
13.8	.31905	.16823	.08871	.04678	.02466	.01301
13.9	.33345	.17938	.09650	.05191	.02792	.01502
14.0	.34827	.19113	.10490	.05757	.03159	.01734
14.1	.36351	.20353	.11396	.06380	.03572	.02000
14.2	.37918	.21659	.12372	.07067	.04037	.02306
14.3	.39528	.23035	.13424	.07823	.04559	.02657
14.4	.41181	.24483	.14556	.08654	.05145	.03059
14.5	.42878	.26007	.15774	.09568	.05803	.03520
14.6	.44619	.27610	.17085	.10572	.06542	.04048
14.7	.46404	.29294	.18493	.11674	.07370	.04653
14.8	.48234	.31065	.20007	.12885	.08299	.05345
14.9	.50108	.32924	.21632	.14214	.09339	.06136
15.0	.52028	.34875	.23378	.15671	.10504	.07041

(table continues)

Table 4-16. Erlang C Delay Loss Probability ($N = 17$ Servers *Continued*)

Offered Traffic (A in Erl)	P (>0)	Delay Loss Probability P (>t) for T_1/T_2 =				
		1.2	1.4	1.6	1.8	2.0
11.1	.07128	.00006	.00002	.00001		
11.2	.07632	.00007	.00002	.00001		
11.3	.08162	.00009	.00003	.00001		
11.4	.08718	.00011	.00003	.00001		
11.5	.09303	.00013	.00004	.00001		
11.6	.09915	.00015	.00005	.00002	.00001	
11.7	.10557	.00018	.00006	.00002	.00001	
11.8	.11228	.00022	.00008	.00003	.00001	
11.9	.11930	.00026	.00009	.00003	.00001	
12.0	.12663	.00031	.00012	.00004	.00002	.00001
12.1	.13428	.00038	.00014	.00005	.00002	.00001
12.2	.14225	.00045	.00017	.00007	.00003	.00001
12.3	.15056	.00053	.00021	.00008	.00003	.00001
12.4	.15920	.00064	.00025	.00010	.00004	.00002
12.5	.16819	.00076	.00031	.00013	.00005	.00002
12.6	.17753	.00090	.00038	.00016	.00006	.00003
12.7	.18723	.00180	.00045	.00019	.00008	.00003
12.8	.19729	.00128	.00055	.00024	.00010	.00004
12.9	.20772	.00152	.00067	.00029	.00013	.00006
13.0	.21853	.00180	.00081	.00036	.00016	.00007
13.1	.22971	.00213	.00098	.00045	.00021	.00009
13.2	.24128	.00252	.00118	.00055	.00026	.00012
13.3	.25324	.00299	.00143	.00068	.00032	.00015
13.4	.26559	.00353	.00172	.00084	.00041	.00020
13.5	.27835	.00417	.00207	.00103	.00051	.00025
13.6	.29151	.00493	.00250	.00127	.00064	.00032
13.7	.30507	.00582	.00301	.00155	.00080	.00042
13.8	.31905	.00686	.00362	.00191	.00101	.00053
13.9	.33345	.00808	.00435	.00234	.00126	.00068
14.0	.34827	.00952	.00522	.00287	.00157	.00086
14.1	.36351	.01120	.00627	.00351	.00197	.00110
14.2	.37918	.01317	.00752	.00430	.00245	.00140
14.3	.39528	.01548	.00902	.00526	.00306	.00179
14.4	.41181	.01818	.01081	.00643	.00382	.00227
14.5	.42878	.02135	.01295	.00785	.00476	.00289
14.6	.44619	.02505	.01550	.00959	.00593	.00367
14.7	.46404	.02937	.01854	.01170	.00739	.00466
14.8	.48234	.03442	.02217	.01428	.00920	.00592
14.9	.50108	.04032	.02649	.01741	.01144	.00751
15.0	.52028	.04720	.03164	.02121	.01422	.00953

Table 4-17. Erlang C Delay Loss Probability (N = 18 Servers)

Offered Traffic (A in Erl)	P (>0)	Delay Loss Probability P (>t) for T_1/T_2 =				
		0.2	0.4	0.6	0.8	1.0
12.1	.08068	.02479	.00762	.00234	.00072	.00022
12.2	.08600	.02696	.00845	.00265	.00083	.00026
12.3	.09157	.02929	.00937	.00300	.00096	.00031
12.4	.09741	.03178	.01037	.00338	.00110	.00036
12.5	.10351	.03446	.01147	.00382	.00127	.00042
12.6	.10989	.03732	.01267	.00430	.00146	.00050
12.7	.11656	.04038	.01399	.00485	.00168	.00058
12.8	.12351	.04366	.01543	.00545	.00193	.00068
12.9	.13076	.04715	.01700	.00613	.00221	.00080
13.0	.13831	.05088	.01872	.00689	.00253	.00093
13.1	.14617	.05486	.02059	.00773	.00290	.00109
13.2	.15434	.05910	.02263	.00866	.00332	.00127
13.3	.16283	.06361	.02485	.00971	.00379	.00148
13.4	.17165	.06841	.02726	.01086	.00433	.00173
13.5	.18080	.07351	.02989	.01215	.00494	.00201
13.6	.19029	.07893	.03274	.01358	.00563	.00234
13.7	.20012	.08468	.03583	.01516	.00642	.00272
13.8	.21030	.09079	.03919	.01692	.00730	.00315
13.9	.22083	.09726	.04284	.01887	.00831	.00366
14.0	.23172	.10412	.04678	.02102	.00945	.00424
14.1	.24297	.11138	.05106	.02341	.01073	.00492
14.2	.25460	.11907	.05568	.02604	.01218	.00570
14.3	.26659	.12719	.06069	.02895	.01381	.00659
14.4	.27896	.13579	.06609	.03217	.01566	.00762
14.5	.29171	.14486	.07194	.03572	.01774	.00881
14.6	.30485	.15444	.07824	.03964	.02008	.01017
14.7	.31838	.16456	.08505	.04396	.02272	.01174
14.8	.33230	.17522	.09239	.04872	.02569	.01355
14.9	.34662	.18646	.10031	.05396	.02903	.01562
15.0	.36134	.19831	.10883	.05973	.03278	.01799
15.1	.37664	.21078	.11802	.06608	.03700	.02071
15.2	.39199	.22391	.12790	.07306	.04173	.02384
15.3	.40792	.23772	.13853	.08073	.04704	.02742
15.4	.42428	.25224	.14996	.08916	.05301	.03151
15.5	.44104	.26751	.16225	.09841	.05969	.03620
15.6	.45823	.28354	.17545	.10857	.06718	.04154
15.7	.47583	.30038	.18963	.11971	.07557	.04771
15.8	.49386	.31806	.20484	.13193	.08497	.05472
15.9	.51231	.33661	.22117	.14532	.09548	.06274
16.0	.53118	.35607	.23868	.15999	.10723	.07189

(table continues)

Table 4-17. Erlang C Delay Loss Probability (*N* = 18 Servers *Continued*)

Offered Traffic (*A* in Erl)	*P* (>0)	Delay Loss Probability *P* (>*t*) for T_1/T_2 =				
		1.2	1.4	1.6	1.8	2.0
12.1	.08068	.00007	.00002	.00001		
12.2	.08600	.00008	.00003	.00001		
12.3	.09157	.00010	.00003	.00001		
12.4	.09741	.00012	.00004	.00001		
12.5	.10351	.00014	.00005	.00002	.00001	
12.6	.10989	.00017	.00006	.00002	.00001	
12.7	.11656	.00020	.00007	.00002	.00001	
12.8	.12351	.00024	.00009	.00003	.00001	
12.9	.13076	.00029	.00010	.00004	.00001	
13.0	.13831	.00034	.00013	.00005	.00002	.00001
13.1	.14617	.00041	.00015	.00006	.00002	.00001
13.2	.15434	.00049	.00019	.00007	.00003	.00001
13.3	.16283	.00058	.00023	.00009	.00003	.00001
13.4	.17165	.00069	.00027	.00011	.00004	.00002
13.5	.18080	.00082	.00033	.00013	.00005	.00002
13.6	.19029	.00097	.00040	.00017	.00007	.00003
13.7	.20012	.00115	.00049	.00021	.00009	.00004
13.8	.21030	.00136	.00059	.00025	.00011	.00005
13.9	.22083	.00161	.00071	.00031	.00014	.00006
14.0	.23172	.00191	.00086	.00039	.00017	.00008
14.1	.24297	.00225	.00103	.00047	.00022	.00010
14.2	.25460	.00266	.00125	.00058	.00027	.00013
14.3	.26659	.00314	.00150	.00072	.00034	.00016
14.4	.27896	.00371	.00181	.00088	.00043	.00021
14.5	.29171	.00437	.00217	.00108	.00054	.00027
14.6	.30485	.00515	.00261	.00132	.00067	.00034
14.7	.31838	.00607	.00314	.00162	.00084	.00043
14.8	.33230	.00714	.00377	.00199	.00105	.00055
14.9	.34662	.00840	.00452	.00243	.00131	.00070
15.0	.36134	.00987	.00542	.00297	.00163	.00090
15.1	.37664	.01160	.00649	.00364	.00204	.00114
15.2	.39199	.01362	.00778	.00444	.00254	.00145
15.3	.40792	.01598	.00931	.00543	.00316	.00184
15.4	.42428	.01874	.01114	.00662	.00394	.00234
15.5	.44104	.02196	.01332	.00808	.00490	.00297
15.6	.45823	.02572	.01592	.00985	.00609	.00377
15.7	.47583	.03012	.01901	.01200	.00758	.00478
15.8	.49386	.03524	.02270	.01462	.00941	.00606
15.9	.51231	.04122	.02708	.01780	.01169	.00768
16.0	.53118	.04819	.03230	.02165	.01451	.00973

Table 4-18. Erlang C Delay Loss Probability (*N* = 19 Servers)

Offered Traffic (*A* in Erl)	*P* (>0)	Delay Loss Probability *P* (>*t*) for T_1/T_2 =				
		0.2	0.4	0.6	0.8	1.0
13.1	.08997	.02765	.00849	.00261	.00080	.00025
13.2	.09553	.02995	.00939	.00294	.00092	.00029
13.3	.10135	.03241	.01037	.00332	.00106	.00034
13.4	.10742	.03505	.01144	.00373	.00122	.00040
13.5	.11375	.03786	.01260	.00420	.00140	.00046
13.6	.12035	.04087	.01388	.00471	.00160	.00054
13.7	.12723	.04408	.01527	.00529	.00183	.00064
13.8	.13439	.04750	.01679	.00593	.00210	.00074
13.9	.14184	.05115	.01844	.00665	.00240	.00086
14.0	.14957	.05502	.02024	.00745	.00274	.00101
14.1	.15761	.05915	.02220	.00833	.00313	.00117
14.2	.16595	.06354	.02433	.00932	.00357	.00137
14.3	.17460	.06820	.02664	.01041	.00407	.00159
14.4	.18356	.07315	.02915	.01162	.00463	.00185
14.5	.19284	.07840	.03188	.01296	.00527	.00214
14.6	.20245	.08397	.03483	.01445	.00599	.00249
14.7	.21239	.08987	.03803	.01609	.00681	.00288
14.8	.22266	.09612	.04150	.01791	.00773	.00334
14.9	.23327	.10274	.04525	.01993	.00878	.00387
15.0	.24422	.10974	.04931	.02216	.00996	.00447
15.1	.25552	.11713	.05369	.02461	.01128	.00517
15.2	.26718	.12495	.05843	.02733	.01278	.00598
15.3	.27919	.13320	.06355	.03032	.01447	.00690
15.4	.29156	.14192	.06908	.03362	.01637	.00797
15.5	.30429	.15111	.07504	.03726	.01850	.00919
15.6	.31740	.16080	.08146	.04127	.02091	.01059
15.7	.33088	.17101	.08839	.04568	.02361	.01220
15.8	.34473	.18177	.09585	.05054	.02665	.01405
15.9	.35896	.19310	.10388	.05588	.03006	.01617
16.0	.37357	.20502	.11252	.06175	.03389	.01860
16.1	.38857	.21756	.12181	.06820	.03819	.02138
16.2	.40396	.23074	.13180	.07529	.04301	.02456
16.3	.41973	.24460	.14254	.08307	.04841	.02821
16.4	.43590	.25915	.15407	.09160	.05446	.03238
16.5	.45247	.27444	.16646	.10096	.06124	.03714
16.6	.46943	.29048	.17974	.11122	.06882	.04259
16.7	.48680	.30731	.19400	.12247	.07731	.04881
16.8	.50457	.32496	.20929	.13479	.08681	.05591
16.9	.52274	.34347	.22567	.14828	.09743	.06401
17.0	.54132	.36286	.24323	.16304	.10929	.07326

(*table continues*)

Table 4-18. Erlang C Delay Loss Probability (*N* = 19 Servers *Continued*)

Offered Traffic (*A* in Erl)	*P* (>0)	Delay Loss Probability *P* (>*t*) for T_1/T_2 =				
		1.2	1.4	1.6	1.8	2.0
13.1	.08997	.00008	.00002	.00001		
13.2	.09553	.00009	.00003	.00001		
13.3	.10135	.00011	.00003	.00001		
13.4	.10742	.00013	.00004	.00001		
13.5	.11375	.00015	.00005	.00002	.00001	
13.6	.12035	.00018	.00006	.00002	.00001	
13.7	.12723	.00022	.00008	.00003	.00001	
13.8	.13439	.00026	.00009	.00003	.00001	
13.9	.14184	.00031	.00011	.00004	.00001	.00001
14.0	.14957	.00037	.00014	.00005	.00002	.00001
14.1	.15761	.00044	.00017	.00006	.00002	.00001
14.2	.16595	.00052	.00020	.00008	.00003	.00001
14.3	.17460	.00062	.00024	.00009	.00004	.00001
14.4	.18356	.00074	.00029	.00012	.00005	.00002
14.5	.19284	.00087	.00035	.00014	.00006	.00002
14.6	.20245	.00103	.00043	.00018	.00007	.00003
14.7	.21239	.00122	.00052	.00022	.00009	.00004
14.8	.22266	.00144	.00062	.00027	.00012	.00005
14.9	.23327	.00170	.00075	.00033	.00015	.00006
15.0	.24422	.00201	.00090	.00041	.00018	.00008
15.1	.25552	.00237	.00109	.00050	.00023	.00010
15.2	.26718	.00280	.00131	.00061	.00029	.00013
15.3	.27919	.00329	.00157	.00075	.00036	.00017
15.4	.29156	.00388	.00189	.00092	.00045	.00022
15.5	.30429	.00456	.00227	.00113	.00056	.00028
15.6	.31740	.00537	.00272	.00138	.00070	.00035
15.7	.33088	.00631	.00326	.00168	.00087	.00045
15.8	.34473	.00741	.00391	.00206	.00109	.00057
15.9	.35896	.00870	.00468	.00252	.00135	.00073
16.0	.37357	.01021	.00560	.00307	.00169	.00093
16.1	.38857	.01197	.00670	.00375	.00210	.00118
16.2	.40396	.01403	.00802	.00458	.00262	.00149
16.3	.41973	.01644	.00958	.00558	.00325	.00190
16.4	.43590	.01925	.01144	.00680	.00404	.00240
16.5	.45247	.02253	.01366	.00829	.00503	.00305
16.6	.46943	.02635	.01631	.01009	.00624	.00386
16.7	.48680	.03081	.01945	.01228	.00775	.00489
16.8	.50457	.03601	.02319	.01494	.00962	.00619
16.9	.52274	.04206	.02764	.01816	.01193	.00784
17.0	.54132	.04911	.03292	.02207	.01479	.00992

Table 4-19. Erlang C Delay Loss Probability (N = 20 Servers)

Offered Traffic (A in Erl)	P (>0)	Delay Loss Probability P (>t) for T_1/T_2 =				
		0.2	0.4	0.6	0.8	1.0
14.1	.09910	.03045	.00936	.00288	.00088	.00027
14.2	.10488	.03288	.01031	.00323	.00101	.00032
14.3	.11091	.03547	.01134	.00363	.00116	.00037
14.4	.11719	.03824	.01248	.00407	.00133	.00043
14.5	.12372	.04118	.01371	.00456	.00152	.00051
14.6	.13052	.04432	.01505	.00511	.00174	.00059
14.7	.13758	.04767	.01651	.00572	.00198	.00069
14.8	.14492	.05122	.01810	.00640	.00226	.00080
14.9	.15253	.05500	.01983	.00715	.00258	.00093
15.0	.16043	.05902	.02171	.00799	.00294	.00108
15.1	.16862	.06328	.02375	.00891	.00335	.00126
15.2	.17710	.06781	.02596	.00994	.00381	.00146
15.3	.18588	.07261	.02836	.01108	.00433	.00169
15.4	.19496	.07769	.03096	.01234	.00492	.00196
15.5	.20435	.08308	.03378	.01373	.00558	.00227
15.6	.21405	.08878	.03683	.01528	.00634	.00263
15.7	.22407	.09482	.04012	.01698	.00718	.00304
15.8	.23441	.10120	.04369	.01886	.00814	.00352
15.9	.24508	.10794	.04754	.02094	.00922	.00406
16.0	.25608	.11506	.05170	.02323	.01044	.00469
16.1	.26741	.12258	.05619	.02576	.01181	.00541
16.2	.27908	.13052	.06104	.02855	.01335	.00624
16.3	.29109	.13889	.06626	.03162	.01508	.00720
16.4	.30345	.14771	.07190	.03500	.01703	.00829
16.5	.31616	.15700	.07796	.03872	.01923	.00955
16.6	.32922	.16679	.08450	.04281	.02169	.01099
16.7	.34264	.17709	.09153	.04731	.02445	.01264
16.8	.35642	.18794	.09910	.05225	.02755	.01453
16.9	.37055	.19934	.10723	.05769	.03103	.01669
17.0	.38506	.21133	.11598	.06365	.03493	.01917
17.1	.39993	.22392	.12537	.07020	.03930	.02201
17.2	.41518	.23715	.13546	.07738	.04420	.02525
17.3	.43079	.25104	.14630	.08525	.04968	.02895
17.4	.44679	.26562	.15792	.09389	.05582	.03318
17.5	.46316	.28092	.17039	.10335	.06268	.03802
17.6	.47991	.29696	.18376	.11371	.07036	.04354
17.7	.49705	.31378	.19808	.12505	.17894	.04983
17.8	.51457	.33140	.21344	.13746	.08853	.05702
17.9	.53248	.34986	.22988	.15104	.09924	.06521
18.0	.55077	.36920	.24748	.16589	.11120	.07454

(table continues)

Table 4-19. Erlang C Delay Loss Probability (N = 20 Servers *Continued*)

Offered Traffic (A in Erl)	P (>0)	Delay Loss Probability P (>t) for T_1/T_2 =				
		1.2	1.4	1.6	1.8	2.0
14.1	.09910	.00008	.00003	.00001		
14.2	.10488	.00010	.00003	.00001		
14.3	.11091	.00012	.00004	.00001		
14.4	.11719	.00014	.00005	.00002		
14.5	.12372	.00017	.00006	.00002	.00001	
14.6	.13052	.00020	.00007	.00002	.00001	
14.7	.13758	.00024	.00008	.00003	.00001	
14.8	.14492	.00028	.00010	.00004	.00001	
14.9	.15253	.00034	.00012	.00004	.00002	.00001
15.0	.16043	.00040	.00015	.00005	.00002	.00001
15.1	.16862	.00047	.00018	.00007	.00002	.00001
15.2	.17710	.00056	.00021	.00008	.00003	.00001
15.3	.18588	.00066	.00026	.00010	.00004	.00002
15.4	.19496	.00078	.00031	.00012	.00005	.00002
15.5	.20435	.00092	.00038	.00015	.00006	.00003
15.6	.21405	.00109	.00045	.00019	.00008	.00003
15.7	.22407	.00129	.00054	.00023	.00010	.00004
15.8	.23441	.00152	.00066	.00028	.00012	.00005
15.9	.24508	.00179	.00079	.00035	.00015	.00007
16.0	.25608	.00211	.00095	.00043	.00019	.00009
16.1	.26741	.00248	.00114	.00052	.00024	.00011
16.2	.27908	.00292	.00137	.00064	.00030	.00014
16.3	.29109	.00343	.00164	.00078	.00037	.00018
16.4	.30345	.00404	.00196	.00096	.00047	.00023
16.5	.31616	.00474	.00235	.00117	.00058	.00029
16.6	.32922	.00557	.00282	.00143	.00072	.00037
16.7	.34264	.00653	.00338	.00174	.00090	.00047
16.8	.35642	.00766	.00404	.00213	.00112	.00059
16.9	.37055	.00898	.00483	.00260	.00140	.00075
17.0	.38506	.01052	.00577	.00317	.00174	.00095
17.1	.39993	.01232	.00690	.00386	.00216	.00121
17.2	.41518	.01442	.00824	.00471	.00269	.00154
17.3	.43079	.01687	.00983	.00573	.00334	.00195
17.4	.44679	.01973	.01173	.00697	.00415	.00246
17.5	.46316	.02306	.01399	.00848	.00515	.00312
17.6	.47991	.02694	.01667	.01032	.00638	.00395
17.7	.49705	.03146	.01986	.01254	.00791	.00500
17.8	.51457	.03672	.02365	.01523	.00981	.00632
17.9	.53248	.04284	.02815	.01850	.01215	.00799
18.0	.55077	.04997	.03349	.02245	.01505	.01009

Table 4-20. Erlang C Delay Loss Probability (*N* = 21 Servers)

Offered Traffic (*A* in Erl)	*P* (>0)	Delay Loss Probability *P* (>*t*) for T_1/T_2 =				
		0.2	0.4	0.6	0.8	1.0
15.1	.10806	.03320	.01020	.00314	.00096	.00030
15.2	.11403	.03575	.01121	.00351	.00110	.00035
15.3	.12025	.03846	.01230	.00393	.00126	.00040
15.4	.12671	.04134	.01349	.00440	.00144	.00047
15.5	.13342	.04441	.01478	.00492	.00164	.00055
15.6	.14038	.04767	.01619	.00550	.00187	.00063
15.7	.14761	.05114	.01772	.00614	.00213	.00074
15.8	.15510	.05482	.01938	.00685	.00242	.00086
15.9	.16286	.05873	.02118	.00764	.00275	.00099
16.0	.17089	.06287	.02313	.00851	.00313	.00115
16.1	.17921	.06726	.02524	.00947	.00356	.00133
16.2	.18781	.07191	.02753	.01054	.00404	.00155
16.3	.19669	.07683	.03001	.01172	.00458	.00179
16.4	.20587	.08204	.03270	.01303	.00519	.00207
16.5	.21535	.08756	.03560	.01447	.00588	.00239
16.6	.22513	.09338	.03873	.01607	.00666	.00276
16.7	.23522	.09954	.04212	.01782	.00754	.00319
16.8	.24561	.10603	.04578	.01976	.00853	.00368
16.9	.25632	.11289	.04972	.02190	.00965	.00425
17.0	.26735	.12013	.05398	.02425	.01090	.00490
17.1	.27870	.12776	.05857	.02685	.01231	.00564
17.2	.29037	.13580	.06351	.02970	.01389	.00650
17.3	.30238	.14427	.06883	.03284	.01567	.00748
17.4	.31471	.15319	.07456	.03629	.01767	.00860
17.5	.32738	.16257	.08073	.04009	.01991	.00989
17.6	.34039	.17245	.08737	.04426	.02242	.01136
17.7	.35374	.18283	.09450	.04884	.02524	.01305
17.8	.36744	.19375	.10216	.05387	.02840	.01498
17.9	.38148	.20522	.11039	.05939	.03195	.01719
18.0	.39587	.21726	.11924	.06544	.03591	.01971
18.1	.41062	.22991	.12872	.07207	.04035	.02259
18.2	.42572	.24318	.13891	.07934	.04532	.02589
18.3	.44118	.25710	.14983	.08731	.05088	.02965
18.4	.45700	.27170	.16153	.09603	.05709	.03394
18.5	.47319	.28700	.17408	.10558	.06404	.03884
18.6	.48974	.30304	.18752	.11603	.07180	.04443
18.7	.50665	.31984	.20191	.12746	.08047	.05080
18.8	.52394	.33744	.21732	.13996	.09014	.05805
18.9	.54159	.35585	.23381	.15363	.10094	.06632
19.0	.55962	.37512	.25145	.16856	.11299	.07574

(*table continues*)

Table 4-20. Erlang C Delay Loss Probability ($N = 21$ Servers *Continued*)

Offered Traffic (A in Erl)	P (>0)	Delay Loss Probability P (>t) for T_1/T_2 =				
		1.2	1.4	1.6	1.8	2.0
15.1	.10806	.00009	.00003	.00001		
15.2	.11403	.00011	.00003	.00001		
15.3	.12025	.00013	.00004	.00001		
15.4	.12671	.00015	.00005	.00002	.00001	
15.5	.13342	.00018	.00006	.00002	.00001	
15.6	.14038	.00022	.00007	.00002	.00001	
15.7	.14761	.00026	.00009	.00003	.00001	
15.8	.15510	.00030	.00011	.00004	.00001	
15.9	.16286	.00036	.00013	.00005	.00002	.00001
16.0	.17089	.00042	.00016	.00006	.00002	.00001
16.1	.17921	.00050	.00019	.00007	.00003	.00001
16.2	.18781	.00059	.00023	.00009	.00003	.00001
16.3	.19669	.00070	.00027	.00011	.00004	.00002
16.4	.20587	.00082	.00033	.00013	.00005	.00002
16.5	.21535	.00097	.00040	.00016	.00007	.00003
16.6	.22513	.00115	.00048	.00020	.00008	.00003
16.7	.23522	.00135	.00057	.00024	.00010	.00004
16.8	.24561	.00159	.00069	.00033	.00013	.00006
16.9	.25632	.00187	.00082	.00036	.00016	.00007
17.0	.26735	.00220	.00099	.00044	.00020	.00009
17.1	.27870	.00259	.00119	.00054	.00025	.00011
17.2	.29037	.00304	.00142	.00066	.00031	.00015
17.3	.30238	.00357	.00170	.00081	.00039	.00018
17.4	.31471	.00419	.00204	.00099	.00048	.00023
17.5	.32738	.00491	.00244	.00121	.00060	.00030
17.6	.34039	.00576	.00292	.00148	.00075	.00038
17.7	.35374	.00674	.00349	.00180	.00093	.00048
17.8	.36744	.00790	.00416	.00220	.00116	.00061
17.9	.38148	.00924	.00497	.00268	.00144	.00077
18.0	.39587	.01082	.00594	.00326	.00179	.00098
18.1	.41062	.01265	.00708	.00397	.00222	.00124
18.2	.42572	.01479	.00845	.00483	.00276	.00157
18.3	.44118	.01728	.01007	.00587	.00342	.00199
18.4	.45700	.02018	.01200	.00713	.00424	.00252
18.5	.47319	.02356	.01429	.00867	.00526	.00319
18.6	.48974	.02749	.01701	.01053	.00651	.00403
18.7	.50665	.03207	.02024	.01278	.00807	.00509
18.8	.52394	.03739	.02408	.01551	.00999	.00643
18.9	.54159	.04358	.02863	.01881	.01236	.00812
19.0	.55962	.05077	.03403	.02281	.01529	.01025

Table 4-21. Erlang C Delay Loss Probability (N = 22 Servers)

Offered Traffic (A in Erl)	P (>0)	Delay Loss Probability P (>t) for T_1/T_2 =				
		0.2	0.4	0.6	0.8	1.0
16.1	.11683	.03590	.01103	.00339	.00104	.00032
16.2	.12297	.03855	.01208	.00379	.00119	.00037
16.3	.12935	.04137	.01323	.00423	.00135	.00043
16.4	.13597	.04436	.01448	.00472	.00154	.00050
16.5	.14283	.04755	.01583	.00527	.00175	.00058
16.6	.14995	.05092	.01729	.00587	.00199	.00068
16.7	.15731	.05450	.01888	.00654	.00227	.00079
16.8	.16494	.05830	.02061	.00728	.00257	.00091
16.9	.17282	.06232	.02247	.00810	.00292	.00105
17.0	.18097	.06658	.02449	.00901	.00331	.00122
17.1	.18940	.07108	.02668	.01001	.00376	.00141
17.2	.19810	.07585	.02904	.01112	.00426	.00163
17.3	.20707	.08089	.03160	.01234	.00482	.00188
17.4	.21634	.08621	.03436	.01369	.00546	.00217
17.5	.22589	.09184	.03734	.01518	.00617	.00251
17.6	.23573	.09778	.04056	.01682	.00698	.00289
17.7	.24586	.10404	.04403	.01863	.00788	.00334
17.8	.25630	.11065	.04777	.02062	.00890	.00384
17.9	.26704	.11761	.05180	.02281	.01005	.00443
18.0	.27808	.12495	.05614	.02523	.01134	.00509
18.1	.28943	.13268	.06082	.02788	.01278	.00586
18.2	.30110	.14081	.06585	.03080	.01440	.00674
18.3	.31308	.14938	.07127	.03400	.01622	.00774
18.4	.32539	.15838	.07709	.03753	.01827	.00889
18.5	.33801	.16785	.08335	.04139	.02055	.01021
18.6	.35096	.17780	.09008	.04564	.02312	.01171
18.7	.36424	.18826	.09730	.05029	.02599	.01343
18.8	.37785	.19924	.10506	.05540	.02921	.01540
18.9	.39180	.21077	.11338	.06099	.03281	.01765
19.0	.40608	.22286	.12231	.06713	.03684	.02022
19.1	.42070	.23555	.13189	.07384	.04134	.02315
19.2	.43567	.24886	.14215	.08120	.04638	.02649
19.3	.45097	.26280	.15315	.08925	.05201	.03031
19.4	.46663	.27742	.16493	.09806	.05830	.03466
19.5	.48263	.29273	.17755	.10769	.06532	.03962
19.6	.49898	.30876	.19106	.11822	.07315	.04527
19.7	.51568	.32554	.20551	.12974	.08190	.05170
19.8	.53274	.34310	.22097	.14231	.09166	.05903
19.9	.55015	.36147	.23751	.15605	.10253	.06737
20.0	.56792	.38069	.25518	.17106	.11466	.07686

(table continues)

Table 4-21. Erlang C Delay Loss Probability ($N = 22$ Servers *Continued*)

Offered Traffic (*A* in Erl)	$P(>0)$	Delay Loss Probability $P(>t)$ for $T_1/T_2 =$				
		1.2	1.4	1.6	1.8	2.0
16.1	.11683	.00010	.00003	.00001		
16.2	.12297	.00012	.00004	.00001		
16.3	.12935	.00014	.00004	.00001		
16.4	.13597	.00016	.00005	.00002	.00001	
16.5	.14283	.00019	.00006	.00002	.00001	
16.6	.14995	.00023	.00008	.00003	.00001	
16.7	.15731	.00027	.00009	.00003	.00001	
16.8	.16494	.00032	.00011	.00004	.00001	.00001
16.9	.17282	.00038	.00014	.00005	.00002	.00001
17.0	.18097	.00045	.00017	.00006	.00002	.00001
17.1	.18940	.00053	.00020	.00007	.00003	.00001
17.2	.19810	.00062	.00024	.00009	.00004	.00001
17.3	.20707	.00074	.00029	.00011	.00004	.00002
17.4	.21634	.00087	.00035	.00014	.00005	.00002
17.5	.22589	.00102	.00041	.00017	.00007	.00003
17.6	.23573	.00120	.00050	.00021	.00009	.00004
17.7	.24586	.00141	.00060	.00025	.00011	.00005
17.8	.25630	.00166	.00072	.00031	.00013	.00006
17.9	.26704	.00195	.00086	.00038	.00017	.00007
18.0	.27808	.00229	.00103	.00046	.00021	.00009
18.1	.28943	.00269	.00123	.00056	.00026	.00012
18.2	.30110	.00315	.00147	.00069	.00032	.00015
18.3	.31308	.00369	.00176	.00084	.00040	.00019
18.4	.32539	.00433	.00211	.00103	.00050	.00024
18.5	.33801	.00507	.00252	.00125	.00062	.00031
18.6	.35096	.00593	.00301	.00152	.00077	.00039
18.7	.36424	.00694	.00359	.00185	.00096	.00050
18.8	.37785	.00812	.00428	.00226	.00119	.00063
18.9	.39180	.00950	.00511	.00275	.00148	.00080
19.0	.40608	.01110	.00609	.00334	.00183	.00101
19.1	.42070	.01296	.00726	.00406	.00227	.00127
19.2	.43567	.01513	.00864	.00494	.00282	.00161
19.3	.45097	.01766	.01029	.00600	.00350	.00204
19.4	.46663	.02061	.01225	.00728	.00433	.00257
19.5	.48263	.02403	.01457	.00884	.00536	.00325
19.6	.49898	.02801	.01733	.01073	.00664	.00411
19.7	.51568	.03264	.02060	.01301	.00821	.00518
19.8	.53274	.03802	.02448	.01577	.01016	.00654
19.9	.55015	.04427	.02908	.01911	.01256	.00825
20.0	.56792	.05152	.03454	.02315	.01552	.01040

Table 4-22. Erlang C Delay Loss Probability (N = 23 Servers)

Offered Traffic (A in Erl)	P (>0)	Delay Loss Probability P (>t) for T_1/T_2 =				
		0.2	0.4	0.6	0.8	1.0
17.1	.12539	.03853	.01184	.00364	.00112	.00034
17.2	.13169	.04128	.01294	.00406	.00127	.00040
17.3	.13822	.04420	.01414	.00452	.00145	.00046
17.4	.14498	.04730	.01543	.00504	.00164	.00054
17.5	.15198	.05059	.01684	.00561	.00187	.00062
17.6	.15922	.05407	.01836	.00624	.00212	.00072
17.7	.16671	.05776	.02001	.00693	.00240	.00083
17.8	.17445	.06166	.02179	.00770	.00272	.00096
17.9	.18244	.06579	.02372	.00855	.00308	.00111
18.0	.19069	.07015	.02581	.00949	.00349	.00128
18.1	.19921	.07477	.02806	.01053	.00395	.00148
18.2	.20799	.07964	.03049	.01168	.00447	.00171
18.3	.21704	.08478	.03312	.01294	.00505	.00197
18.4	.22637	.09021	.03595	.01433	.00571	.00228
18.5	.23598	.09594	.03901	.01586	.00645	.00262
18.6	.24587	.10198	.04230	.01755	.00728	.00302
18.7	.25604	.10835	.04585	.01940	.00821	.00347
18.8	.26650	.11505	.04967	.02144	.00926	.00400
18.9	.27726	.12211	.05378	.02369	.01043	.00459
19.0	.28831	.12955	.05821	.02615	.01175	.00528
19.1	.29966	.13736	.06297	.02887	.01323	.00607
19.2	.31131	.14559	.06809	.03184	.01489	.00696
19.3	.32326	.15423	.07359	.03511	.01675	.00799
19.4	.33553	.16332	.07950	.03870	.01884	.00917
19.5	.34810	.17286	.08584	.04263	.02117	.01051
19.6	.36099	.18289	.09265	.04694	.02378	.01205
19.7	.37420	.19341	.09996	.05167	.02670	.01380
19.8	.38772	.20444	.10780	.05684	.02997	.01580
19.9	.40157	.21602	.11621	.06251	.03363	.01809
20.0	.41574	.22816	.12522	.06872	.03772	.02070
20.1	.43024	.24089	.13487	.07552	.04228	.02367
20.2	.44506	.25422	.14522	.08295	.04738	.02706
20.3	.46022	.26819	.15629	.09108	.05308	.03093
20.4	.47571	.28282	.16814	.09996	.05943	.03533
20.5	.49153	.29813	.18082	.10968	.06652	.04035
20.6	.50769	.31415	.19439	.12029	.07443	.04606
20.7	.52419	.33091	.20890	.13188	.08325	.05256
20.8	.54103	.34844	.22441	.14453	.09308	.05995
20.9	.55821	.36677	.24099	.15834	.10404	.06836
21.0	.57573	.38593	.25869	.17341	.11624	.07792

(table continues)

Table 4-22. Erlang C Delay Loss Probability ($N = 23$ Servers *Continued*)

Offered Traffic (A in Erl)	$P(>0)$	Delay Loss Probability $P(>t)$ for $T_1/T_2 =$				
		1.2	1.4	1.6	1.8	2.0
17.1	.12539	.00011	.00003	.00001		
17.2	.13169	.00012	.00004	.00001		
17.3	.13822	.00015	.00005	.00002		
17.4	.14498	.00017	.00006	.00002	.00001	
17.5	.15198	.00021	.00007	.00002	.00001	
17.6	.15922	.00024	.00008	.00003	.00001	
17.7	.16671	.00029	.00010	.00003	.00001	
17.8	.17445	.00034	.00012	.00004	.00002	.00001
17.9	.18244	.00040	.00014	.00005	.00002	.00001
18.0	.19069	.00047	.00017	.00006	.00002	.00001
18.1	.19921	.00056	.00021	.00008	.00003	.00001
18.2	.20799	.00066	.00025	.00010	.00004	.00001
18.3	.21704	.00077	.00030	.00012	.00005	.00002
18.4	.22637	.00091	.00036	.00014	.00006	.00002
18.5	.23598	.00107	.00043	.00018	.00007	.00003
18.6	.24587	.00125	.00052	.00022	.00009	.00004
18.7	.25604	.00147	.00062	.00026	.00011	.00005
18.8	.26650	.00173	.00074	.00032	.00014	.00006
18.9	.27726	.00202	.00089	.00039	.00017	.00008
19.0	.28831	.00237	.00107	.00048	.00022	.00010
19.1	.29966	.00278	.00127	.00058	.00027	.00012
19.2	.31131	.00326	.00152	.00071	.00033	.00016
19.3	.32326	.00381	.00182	.00087	.00041	.00020
19.4	.33553	.00446	.00217	.00106	.00051	.00025
19.5	.34810	.00522	.00259	.00129	.00064	.00032
19.6	.36099	.00610	.00309	.00157	.00079	.00040
19.7	.37420	.00713	.00369	.00191	.00098	.00051
19.8	.38772	.00833	.00439	.00232	.00122	.00064
19.9	.40157	.00973	.00524	.00282	.00151	.00081
20.0	.41574	.01136	.00623	.00342	.00188	.00103
20.1	.43024	.01325	.00742	.00416	.00233	.00130
20.2	.44506	.01546	.00883	.00504	.00288	.00165
20.3	.46022	.01802	.01050	.00612	.00357	.00208
20.4	.47571	.02101	.01249	.00742	.00441	.00262
20.5	.49153	.02447	.01484	.00900	.00546	.00331
20.6	.50769	.02850	.01764	.01091	.00675	.00418
20.7	.52419	.03318	.02094	.01322	.00835	.00527
20.8	.54103	.03861	.02487	.01601	.01031	.00664
20.9	.55821	.04491	.02951	.01939	.01274	.00837
21.0	.57573	.05223	.03501	.02347	.01573	.01055

Table 4-23. Erlang C Delay Loss Probability ($N = 24$ Servers)

Offered Traffic (A in Erl)	$P\,(>0)$	Delay Loss Probability $P\,(>t)$ for $T_1/T_2 =$				
		0.2	0.4	0.6	0.8	1.0
18.1	.13375	.04110	.01263	.00388	.00119	.00037
18.2	.14019	.04395	.01378	.00432	.00135	.00042
18.3	.14684	.04696	.01502	.00480	.00154	.00049
18.4	.15373	.05016	.01637	.00534	.00174	.00057
18.5	.16085	.05354	.01782	.00593	.00197	.00066
18.6	.16820	.05712	.01940	.00659	.00224	.00076
18.7	.17580	.06091	.02110	.00731	.00253	.00088
18.8	.18364	.06491	.02294	.00811	.00287	.00101
18.9	.19172	.06914	.02493	.00899	.00324	.00117
19.0	.20006	.07360	.02708	.00996	.00366	.00135
19.1	.20866	.07831	.02939	.01103	.00414	.00155
19.2	.21751	.08328	.03189	.01221	.00468	.00179
19.3	.22663	.08853	.03458	.01351	.00528	.00206
19.4	.23601	.09405	.03748	.01494	.00595	.00237
19.5	.24566	.09988	.04061	.01651	.00671	.00273
19.6	.25558	.10601	.04397	.01824	.00757	.00314
19.7	.26578	.11247	.04759	.02104	.00852	.00361
19.8	.27626	.11927	.05149	.02223	.00960	.00414
19.9	.28703	.12642	.05568	.02452	.01080	.00476
20.0	.29807	.13393	.06018	.02704	.01215	.00546
20.1	.30941	.14184	.06502	.02980	.01366	.00626
20.2	.32104	.15014	.07022	.03284	.01536	.00718
20.3	.33296	.15886	.07580	.03616	.01725	.00823
20.4	.34518	.16802	.08178	.03981	.01938	.00943
20.5	.35770	.17763	.08821	.04380	.02175	.01080
20.6	.37053	.18772	.09510	.04818	.02441	.01237
20.7	.38366	.19829	.10249	.05297	.02738	.01415
20.8	.39709	.20938	.11041	.05822	.03070	.01619
20.9	.41084	.22101	.11889	.06396	.03441	.01851
21.0	.42490	.23319	.12798	.07024	.03855	.02115
21.1	.43927	.24595	.13771	.07710	.04317	.02417
21.2	.45396	.25931	.14812	.08461	.04833	.02761
21.3	.46897	.27329	.15926	.09281	.05408	.03152
21.4	.48430	.28792	.17118	.10177	.06050	.03597
21.5	.49995	.30323	.18392	.11155	.06766	.04104
21.6	.51592	.31925	.19754	.12224	.07564	.04680
21.7	.53223	.33599	.21210	.13390	.08453	.05336
21.8	.54886	.35348	.22766	.14662	.09443	.06082
21.9	.56581	.37177	.24427	.16050	.15045	.06929
22.0	.58310	.39087	.26201	.17563	.11773	.07892

(*table continues*)

Table 4-23. Erlang C Delay Loss Probability ($N = 24$ Servers *Continued*)

Offered Traffic (*A* in Erl)	$P(>0)$	Delay Loss Probability $P(>t)$ for $T_1/T_2 =$				
		1.2	1.4	1.6	1.8	2.0
18.1	.13375	.00011	.00003	.00001		
18.2	.14019	.00013	.00004	.00001		
18.3	.14684	.00016	.00005	.00002	.00001	
18.4	.15373	.00019	.00006	.00002	.00001	
18.5	.16085	.00022	.00007	.00002	.00001	
18.6	.16820	.00026	.00009	.00003	.00001	
18.7	.17580	.00030	.00011	.00004	.00001	
18.8	.18364	.00036	.00013	.00004	.00002	.00001
18.9	.19172	.00042	.00015	.00005	.00002	.00001
19.0	.20006	.00050	.00018	.00007	.00002	.00001
19.1	.20866	.00058	.00022	.00008	.00003	.00001
19.2	.21751	.00069	.00026	.00010	.00004	.00001
19.3	.22663	.00081	.00031	.00012	.00005	.00002
19.4	.23601	.00095	.00038	.00015	.00006	.00002
19.5	.24566	.00111	.00045	.00018	.00007	.00003
19.6	.25558	.00130	.00054	.00022	.00009	.00004
19.7	.26578	.00153	.00065	.00027	.00012	.00005
19.8	.27626	.00179	.00077	.00033	.00014	.00006
19.9	.28703	.00210	.00092	.00041	.00018	.00008
20.0	.29807	.00245	.00110	.00050	.00022	.00010
20.1	.30941	.00287	.00132	.00060	.00028	.00013
20.2	.32104	.00336	.00157	.00073	.00034	.00016
20.3	.33296	.00393	.00187	.00089	.00043	.00020
20.4	.34518	.00459	.00223	.00109	.00053	.00026
20.5	.35770	.00536	.00266	.00132	.00066	.00033
20.6	.37053	.00626	.00317	.00161	.00081	.00041
20.7	.38366	.00731	.00378	.00195	.00101	.00052
20.8	.39709	.00854	.00450	.00237	.00125	.00066
20.9	.41084	.00996	.00536	.00288	.00155	.00083
21.0	.42490	.01161	.00637	.00350	.00192	.00105
21.1	.43927	.01351	.00758	.00424	.00238	.00133
21.2	.45396	.01577	.00901	.00515	.00294	.00168
21.3	.46897	.01837	.01070	.00624	.00363	.00212
21.4	.48430	.02139	.01271	.00756	.00449	.00267
21.5	.49995	.02489	.01510	.00916	.00555	.00337
21.6	.51592	.02896	.01792	.01109	.00686	.00425
21.7	.53223	.03369	.02127	.01342	.00847	.00535
21.8	.54886	.03917	.02523	.01625	.01046	.00674
21.9	.56581	.04553	.02991	.01965	.01291	.00849
22.0	.58310	.05290	.03456	.02377	.01593	.01068

Table 4-24. Erlang C Delay Loss Probability ($N = 25$ Servers)

Offered Traffic (A in Erl)	P (>0)	Delay Loss Probability P (>t) for T_1/T_2 =				
		0.2	0.4	0.6	0.8	1.0
19.1	.14191	.04360	.01340	.00412	.00127	.00039
19.2	.14846	.04654	.01459	.00457	.00143	.00045
19.3	.15524	.04965	.01588	.00508	.00162	.00052
19.4	.16223	.05293	.01727	.00564	.00184	.00060
19.5	.16946	.05641	.01878	.00625	.00208	.00069
19.6	.17691	.06008	.02040	.00693	.00235	.00080
19.7	.18460	.06396	.02216	.00768	.00266	.00092
19.8	.19252	.06805	.02405	.00850	.00300	.00106
19.9	.20069	.07237	.02610	.00941	.00339	.00122
20.0	.20910	.07692	.02830	.01041	.00383	.00141
20.1	.21776	.08173	.03067	.01151	.00432	.00162
20.2	.22667	.08679	.03323	.01272	.00487	.00187
20.3	.23584	.09213	.03599	.01406	.00549	.00215
20.4	.24527	.09774	.03895	.01552	.00619	.00247
20.5	.25495	.10366	.04214	.01713	.00697	.00283
20.6	.26490	.10988	.04558	.01890	.00784	.00325
20.7	.27512	.11642	.04926	.02085	.00882	.00373
20.8	.28561	.12330	.05323	.02298	.00992	.00428
20.9	.29637	.13053	.05749	.02532	.01115	.00491
21.0	.30741	.13813	.06207	.02789	.01253	.00563
21.1	.31873	.14611	.06698	.03070	.01407	.00645
21.2	.33033	.15448	.07225	.03379	.01580	.00739
21.3	.34221	.16328	.07790	.03717	.01773	.00846
21.4	.35439	.17250	.08396	.04087	.01989	.00968
21.5	.36685	.18217	.09046	.04492	.02231	.01108
21.6	.37960	.19231	.09743	.04936	.02501	.01267
21.7	.39265	.20294	.10489	.05421	.02802	.01448
21.8	.40600	.21408	.11288	.05952	.03139	.01655
21.9	.41965	.22575	.12144	.06533	.03514	.01891
22.0	.43359	.23796	.13060	.07167	.03934	.02159
22.1	.44781	.25075	.14039	.07861	.04401	.02464
22.2	.46240	.26413	.15087	.08618	.04923	.02812
22.3	.47727	.27813	.16208	.09445	.05504	.03208
22.4	.49244	.29277	.17406	.10348	.06152	.03658
22.5	.50793	.30807	.18686	.11333	.06874	.04169
22.6	.52372	.32407	.20053	.12409	.07678	.04751
22.7	.53984	.34079	.21514	.13581	.08574	.05412
22.8	.55626	.35826	.23073	.14860	.09570	.06164
22.9	.57301	.37650	.24738	.16254	.10680	.07017
23.0	.59008	.39554	.26514	.17773	.11914	.07986

(*table continues*)

Table 4-24. Erlang C Delay Loss Probability ($N = 25$ Servers *Continued*)

Offered Traffic (*A* in Erl)	$P (>0)$	Delay Loss Probability $P (>t)$ for $T_1/T_2 =$				
		1.2	1.4	1.6	1.8	2.0
19.1	.14191	.00012	.00004	.00001		
19.2	.14846	.00014	.00004	.00001		
19.3	.15524	.00017	.00005	.00002	.00001	
19.4	.16223	.00020	.00006	.00002	.00001	
19.5	.16946	.00023	.00008	.00003	.00001	
19.6	.17691	.00027	.00009	.00003	.00001	
19.7	.18460	.00032	.00011	.00004	.00001	
19.8	.19252	.00038	.00013	.00005	.00002	.00001
19.9	.20069	.00044	.00016	.00006	.00002	.00001
20.0	.20910	.00052	.00019	.00007	.00003	.00001
20.1	.21776	.00061	.00023	.00009	.00003	.00001
20.2	.22667	.00071	.00027	.00010	.00004	.00002
20.3	.23584	.00084	.00033	.00013	.00005	.00002
20.4	.24527	.00098	.00039	.00016	.00006	.00002
20.5	.25495	.00115	.00047	.00019	.00008	.00003
20.6	.26490	.00135	.00056	.00023	.00010	.00004
20.7	.27512	.00158	.00067	.00028	.00012	.00005
20.8	.28561	.00185	.00080	.00034	.00015	.00006
20.9	.29637	.00216	.00095	.00042	.00018	.00008
21.0	.30741	.00253	.00114	.00051	.00023	.00010
21.1	.31873	.00296	.00136	.00062	.00028	.00013
21.2	.33033	.00346	.00162	.00076	.00035	.00017
21.3	.34221	.00404	.00193	.00092	.00044	.00021
21.4	.35439	.00471	.00229	.00112	.00054	.00026
21.5	.36685	.00550	.00273	.00136	.00067	.00033
21.6	.37960	.00642	.00325	.00165	.00083	.00042
21.7	.39265	.00749	.00387	.00200	.00103	.00053
21.8	.40600	.00873	.00460	.00243	.00128	.00067
21.9	.41965	.01017	.00547	.00294	.00158	.00085
22.0	.43359	.01185	.00650	.00357	.00196	.00107
22.1	.44781	.01380	.00773	.00433	.00242	.00136
22.2	.46240	.01606	.00917	.00524	.00299	.00171
22.3	.47727	.01869	.01089	.00635	.00370	.00216
22.4	.49244	.02175	.01293	.00769	.00457	.00272
22.5	.50793	.02529	.01534	.00930	.00564	.00342
22.6	.52372	.02940	.01819	.01126	.00697	.00431
22.7	.53984	.03417	.02157	.01362	.00860	.00543
22.8	.55626	.03970	.02557	.01647	.01060	.00683
22.9	.57301	.04611	.03029	.01990	.01308	.00859
23.0	.59008	.05353	.03588	.02405	.01612	.01081

5

Binomial Distribution

The Binomial distribution is used in lieu of the Poisson distribution for applications such as final trunk groups between a small PBX or remote switching unit and a central office (CO). It is based on the following assumptions:

- Calls are served in random order.
- There are a finite number of sources.
- Blocked calls are held.
- Holding time is exponential or constant.

5.1   BINOMIAL FORMULA

The Binomial formula is given in Equation 5.1.

$$P = \sum_{i=N}^{S-1} \frac{(S-1)!}{i! \, (S-1-i)!} \, a^i \, (1-a)^{(S-1-i)} \qquad (5.1)$$

where P = Binomial loss probability
S = Number of sources

131

N = Number of trunks in full-availability group

a = Traffic offered to group in Erlangs per source

5.2   BINOMIAL COMPUTER PROGRAM

The following computer program can be used to calculate Equation 5.1 to determine Binomial loss probabilities. Required inputs are the number of sources, the number of servers in the group, and the total traffic offered to the group expressed in Erlangs (Erlangs are converted to Erlangs per source in Step 140).

```
100 REM BINOMIAL LOSS PROBABILITY CALCULATION
110 INPUT "ENTER NUMBER OF SOURCES (S)";S
120 INPUT "ENTER NUMBER OF SERVERS (N)";N
130 INPUT "ENTER OFFERED TRAFFIC IN ERLANGS (A)";A
140 LET A=A/S
150 FOR I=N TO (S-1)
160 LET X=1
170 FOR J=1 TO I
180 LET X=X*(S-J)/J
190 NEXT J
200 LET P=P+X*A^I*(1-A)^(S-I-1)
210 NEXT I
220 PRINT USING "P = #.#####";P
230 END
```

5.3   BINOMIAL LOSS PROBABILITY TABLES

Binomial loss probability tables (Tables 5-1 through 5-6) are used to determine the loss probability (grade of service) when traffic is offered to a group of trunks from a finite group of sources. Blanks (no data) in the tables indicate that the loss probability is less than 0.000005—not necessarily zero, but essentially nonblocking for practical applications. The following examples illustrate typical table usage.

Example 5-1

Determine the loss probability for a 60-line digital remote switching unit connected to the local CO by a 24-channel trunk group, given that nonlocal busy-hour traffic is 15 Erlangs.

$$a = \frac{(15 \text{ Erl})}{60 \text{ lines}} = 0.25 \text{ Erl per line (source)}$$

Using Table 5-6 (60 sources) (pages 144–145), select the a row for 0.25 Erlangs, the P column for 24 servers, and read .00587 at the intersection.

Example 5-2

For the remote switching unit of Example 5-1, determine the offered traffic in Erlangs and CCS to achieve a grade of service of 0.01 or better.

Using Table 5-6 (60 sources), select the N column for 24 servers and read down until .00979 is found. Read across the a row to determine that 0.26 Erlangs per source (15.6 Erlangs) of traffic can be offered.

$$(15.6 \text{ Erl})(36 \text{ CCS/Erl}) = 561.6 \text{ CCS}$$

Example 5-3

Determine the number of CO trunks required for a 30-line PBX in a small office if nonlocal busy-hour traffic is 0.25 Erlangs per subscriber and a grade of service of 0.05 or better is required.

Using Table 5-3 (30 sources) (pages 137–138), select the a row for 0.25 Erlangs per source and read across until .03903 is found. Read up the N column to determine that at least 12 trunks are required.

Table 5-1. Binomial Loss Probability (S = 10 Sources)

Traffic per Source (a in Erl)	Loss Probability (P) for N =					
	3	4	5	6	7	8
0.01	.00008					
0.02	.00061	.00002				
0.03	.00198	.00009				
0.04	.00448	.00027	.00001			
0.05	.00836	.00064	.00003			
0.06	.01380	.00128	.00008			
0.07	.02091	.00227	.00017	.00001		
0.08	.02979	.00372	.00031	.00002		
0.09	.04048	.00570	.00055	.00004		
0.10	.05297	.00833	.00089	.00006		
0.11	.06725	.01169	.00138	.00011	.00001	
0.12	.08326	.01585	.00206	.00018	.00001	
0.13	.10093	.02094	.00296	.00028	.00002	
0.14	.12016	.02691	.00414	.00043	.00003	
0.15	.14085	.03393	.00563	.00063	.00005	
0.16	.16289	.04202	.00748	.00091	.00007	
0.17	.18614	.05121	.00976	.00127	.00011	.00001
0.18	.21046	.06153	.01250	.00173	.00016	.00001
0.19	.23573	.07301	.01575	.00232	.00022	.00001
0.20	.26180	.08564	.01958	.00307	.00031	.00002
0.21	.28853	.09943	.02403	.00398	.00043	.00003
0.22	.31579	.11436	.02914	.00511	.00059	.00004
0.23	.34342	.13041	.03497	.00646	.00078	.00006
0.24	.37131	.14754	.04155	.00808	.00103	.00008
0.25	.39932	.16573	.04893	.00999	.00134	.00011
0.26	.42733	.18490	.05713	.01224	.00173	.00014
0.27	.45523	.20502	.06620	.01486	.00220	.00019
0.28	.48290	.22601	.07616	.01788	.00277	.00026
0.29	.51025	.24781	.08702	.02135	.00346	.00033
0.30	.53717	.27034	.09881	.02529	.00429	.00043
0.31	.56358	.29352	.11153	.02976	.00527	.00056
0.32	.58941	.31728	.12519	.03479	.00643	.00071
0.33	.61459	.34152	.13978	.04041	.00778	.00089
0.34	.63904	.36916	.15529	.04666	.00936	.00112
0.35	.66273	.39111	.17172	.05359	.01118	.00140
0.36	.68559	.41627	.18904	.06121	.01328	.00173
0.37	.70760	.44157	.20722	.06958	.01569	.00212
0.38	.72872	.46692	.22622	.07870	.01842	.00259
0.39	.74893	.49222	.24602	.08862	.02153	.00315
0.40	.76821	.51739	.26657	.09935	.02503	.00380

Table 5-2. Binomial Loss Probability (S = 20 Sources)

Traffic per Source (a in Erl)	Loss Probability (P) for N =					
	3	4	5	6	7	8
0.01	.00086	.00003				
0.02	.00610	.00049	.00003			
0.03	.01826	.00219	.00020	.00001		
0.04	.03840	.00612	.00074	.00007	.00001	
0.05	.06655	.01324	.00201	.00024	.00002	
0.06	.10207	.02430	.00444	.00064	.00007	.00001
0.07	.14392	.03985	.00851	.00144	.00020	.00002
0.08	.19084	.06016	.01471	.00285	.00045	.00006
0.09	.24148	.08527	.02347	.00514	.00091	.00013
0.10	.29456	.11500	.03519	.00859	.00170	.00027
0.11	.34883	.14897	.05016	.01352	.00295	.00053
0.12	.40324	.18667	.06854	.02022	.00484	.00095
0.13	.45683	.22750	.09043	.02899	.00756	.00162
0.14	.50885	.27079	.11578	.04007	.01132	.00262
0.15	.55868	.31585	.14444	.05370	.01633	.00408
0.16	.60585	.36199	.17618	.07001	.02282	.00613
0.17	.65004	.40854	.21068	.08911	.03101	.00891
0.18	.69104	.45490	.24756	.11102	.04108	.01257
0.19	.72873	.50052	.28639	.13571	.05323	.01731
0.20	.76311	.54491	.32671	.16306	.06760	.02328
0.21	.79422	.58767	.36806	.19292	.08429	.03066
0.22	.82217	.62848	.40996	.22505	.10337	.03962
0.23	.84710	.66706	.45196	.25920	.12485	.05032
0.24	.86920	.70325	.49363	.29504	.14871	.06290
0.25	.88866	.73691	.53458	.33224	.17488	.07746
0.26	.90569	.76798	.57445	.37046	.20322	.09409
0.27	.92051	.79646	.61294	.40931	.23357	.11286
0.28	.93332	.82238	.64979	.44844	.26573	.13378
0.29	.94435	.84580	.68479	.48750	.29947	.15683
0.30	.95378	.86683	.71778	.52614	.33450	.18197
0.31	.96180	.88559	.74864	.56405	.37055	.20909
0.32	.96858	.90222	.77730	.60095	.40731	.23808
0.33	.97429	.91687	.80374	.63659	.44448	.26876
0.34	.97907	.92970	.82797	.67074	.48176	.30095
0.35	.98304	.94086	.85000	.70324	.51883	.33443
0.36	.98634	.95052	.86992	.73392	.55543	.36896
0.37	.98905	.95882	.88781	.76271	.59126	.40427
0.38	.99127	.96592	.90377	.78951	.62610	.44010
0.39	.99307	.97179	.91792	.81430	.65972	.47618
0.40	.99454	.97704	.93039	.83708	.69193	.51222

(*table continues*)

Table 5-2. Binomial Loss Probability (S = 20 Sources *Continued*)

Traffic per Source (*a* in Erl)	Loss Probability (*P*) for *N* =					
	9	10	11	12	13	14
0.01						
0.02						
0.03						
0.04						
0.05						
0.06						
0.07						
0.08	.00001					
0.09	.00002					
0.10	.00004					
0.11	.00008	.00001				
0.12	.00015	.00002				
0.13	.00028	.00004				
0.14	.00050	.00008	.00001			
0.15	.00084	.00014	.00002			
0.16	.00136	.00025	.00004			
0.17	.00212	.00042	.00007	.00001		
0.18	.00319	.00067	.00012	.00002		
0.19	.00467	.00104	.00019	.00003		
0.20	.00666	.00158	.00031	.00005		
0.21	.00928	.00233	.00048	.00008	.00001	
0.22	.01266	.00336	.00074	.00013	.00002	
0.23	.01693	.00474	.00110	.00021	.00003	
0.24	.02225	.00656	.00160	.00032	.00005	.00001
0.25	.02875	.00890	.00229	.00048	.00008	.00001
0.26	.03658	.01189	.00321	.00071	.00013	.00002
0.27	.04589	.01562	.00442	.00103	.00020	.00003
0.28	.05681	.02022	.00599	.00147	.00029	.00005
0.29	.06945	.02582	.00800	.00205	.00043	.00007
0.30	.08392	.03255	.01054	.00282	.00062	.00011
0.31	.10029	.04054	.01370	.00383	.00088	.00016
0.32	.11862	.04991	.01758	.00513	.00122	.00023
0.33	.13894	.06079	.02229	.00678	.00169	.00034
0.34	.16124	.07328	.02796	.00886	.00230	.00048
0.35	.18549	.08747	.03469	.01144	.00309	.00067
0.36	.21163	.10346	.04262	.01461	.00411	.00093
0.37	.23954	.12129	.05184	.01847	.00541	.00128
0.38	.26910	.14101	.06249	.02312	.00704	.00173
0.39	.30016	.16261	.07467	.02867	.00906	.00231
0.40	.33252	.18609	.08847	.03523	.01156	.00307

Table 5-3. Binomial Loss Probability (S = 30 Sources)

Traffic per Source (a in Erl)	Loss Probability (P) for N =					
	9	10	11	12	13	14
0.11	.00301	.00072	.00015	.00003		
0.12	.00542	.00141	.00032	.00006	.00001	
0.13	.00916	.00260	.00065	.00014	.00003	
0.14	.01465	.00451	.00121	.00029	.00006	.00001
0.15	.02234	.00742	.00215	.00055	.00012	.00002
0.16	.03270	.01165	.00363	.00099	.00024	.00005
0.17	.04616	.01757	.00586	.00171	.00044	.00010
0.18	.06309	.02556	.00908	.00284	.00078	.00019
0.19	.08377	.03600	.01359	.00451	.00132	.00034
0.20	.10838	.04926	.01970	.00694	.00215	.00059
0.21	.13697	.06566	.02775	.01034	.00340	.00098
0.22	.16946	.08544	.03805	.01496	.00519	.00159
0.23	.20561	.10878	.05094	.02109	.00772	.00249
0.24	.24510	.13575	.06669	.02901	.01117	.00380
0.25	.28746	.16631	.08554	.03903	.01578	.00565
0.26	.33214	.20030	.10765	.05143	.02180	.00818
0.27	.37852	.23747	.13312	.06647	.02984	.01160
0.28	.42595	.27745	.16195	.08437	.03911	.01610
0.29	.47376	.31980	.19403	.10530	.05094	.02191
0.30	.52130	.36400	.22918	.12938	.06522	.02927
0.31	.56794	.40949	.26711	.15663	.08217	.03843
0.32	.61312	.45566	.30746	.18699	.10196	.04963
0.33	.65636	.50191	.34977	.22034	.12472	.06313
0.34	.69724	.54767	.39357	.25645	.15050	.07912
0.35	.73545	.59239	.43832	.29502	.17928	.09779
0.36	.77076	.63556	.48346	.33568	.21099	.11928
0.37	.80302	.67675	.52844	.37799	.24546	.14366
0.38	.83218	.71562	.57274	.42149	.28243	.17098
0.39	.85825	.75187	.61586	.46565	.32160	.20116
0.40	.88131	.78532	.65733	.50996	.36258	.23410
0.41	.90149	.81584	.69679	.55390	.40496	.26961
0.42	.91897	.84339	.73391	.59699	.44826	.30742
0.43	.93396	.86799	.76845	.63875	.49198	.34719
0.44	.94667	.88972	.80023	.67878	.53563	.38856
0.45	.95733	.90871	.82915	.71672	.57873	.43109
0.46	.96619	.92514	.85520	.75229	.62079	.47430
0.47	.97347	.93919	.87839	.78526	.66139	.51773
0.48	.97939	.95109	.89883	.81551	.7 0014	.56088
0.49	.98415	.96104	.91663	.84293	.73672	.60327
0.50	.98794	.96929	.93198	.86753	.77087	.64446

(table continues)

Table 5-3. Binomial Loss Probability (S = 30 Sources *Continued*)

Traffic per Source (*a* in Erl)	Loss Probability (P) for $N =$					
	15	16	17	18	19	20
0.11						
0.12						
0.13						
0.14						
0.15						
0.16	.00001					
0.17	.00002					
0.18	.00004	.00001				
0.19	.00008	.00002				
0.20	.00014	.00003	.00001			
0.21	.00025	.00006	.00001			
0.22	.00043	.00010	.00002			
0.23	.00071	.00018	.00004	.00001		
0.24	.00114	.00030	.00007	.00001		
0.25	.00178	.00050	.00012	.00003		
0.26	.00272	.00080	.00020	.00005	.00001	
0.27	.00404	.00124	.00034	.00008	.00002	
0.28	.00587	.00189	.00054	.00013	.00003	.00001
0.29	.00835	.00281	.00084	.00022	.00005	.00001
0.30	.01165	.00411	.00128	.00035	.00008	.00002
0.31	.01596	.00587	.00191	.00054	.00014	.00003
0.32	.02149	.00825	.00279	.00083	.00022	.00005
0.33	.02846	.01138	.00402	.00125	.00034	.00008
0.34	.03709	.01545	.00569	.00184	.00052	.00013
0.35	.04764	.02063	.00791	.00267	.00079	.00020
0.36	.06031	.02715	.01083	.00380	.00117	.00031
0.37	.07534	.03522	.01460	.00534	.00171	.00048
0.38	.09291	.04506	.01940	.00737	.00246	.00071
0.39	.11316	.05690	.02542	.01004	.00348	.00105
0.40	.13621	.07095	.03288	.01348	.00485	.00152
0.41	.16211	.08743	.04119	.01786	.00667	.00218
0.42	.19086	.10645	.05297	.02336	.00906	.00307
0.43	.22237	.12820	.06604	.03018	.01215	.00427
0.44	.25650	.15273	.08139	.03853	.01608	.00586
0.45	.29303	.18008	.09922	.04862	.02103	.00795
0.46	.33169	.21021	.11966	.06068	.02718	.01066
0.47	.37214	.24303	.14284	.07491	.03474	.01412
0.48	.41396	.27835	.16882	.09150	.04392	.01850
0.49	.45674	.31596	.19760	.11064	.04495	.02396
0.50	.50000	.35554	.22913	.13247	.06802	.03071

Table 5-4. Binomial Loss Probability (S = 40 Sources)

Traffic per Source (a in Erl)	Loss Probability (P) for N =					
	13	14	15	16	17	18
0.11	.00017	.00004	.00001			
0.12	.00041	.00010	.00002			
0.13	.00090	.00024	.00006	.00001		
0.14	.00180	.00052	.00014	.00003	.00001	
0.15	.00337	.00106	.00030	.00008	.00002	
0.16	.00593	.00200	.00061	.00004	.00017	.00001
0.17	.00991	.00358	.00117	.00035	.00009	.00002
0.18	.01591	.00608	.00212	.00067	.00019	.00005
0.19	.02413	.00987	.00366	.00123	.00038	.00010
0.20	.03550	.01539	.00605	.00216	.00070	.00021
0.21	.05045	.02311	.00961	.00363	.00125	.00039
0.22	.06950	.03356	.01473	.00588	.00213	.00070
0.23	.09305	.04724	.02183	.00918	.00351	.00122
0.24	.12134	.06464	.03139	.01388	.00559	.00205
0.25	.15447	.08615	.04387	.02037	.00862	.00333
0.26	.19231	.11208	.05972	.02907	.01291	.00523
0.27	.23456	.14256	.07936	.04040	.01879	.00798
0.28	.28071	.17758	.10310	.05482	.02666	.01184
0.29	.33008	.21695	.13113	.07271	.03692	.01714
0.30	.38185	.26028	.16352	.09441	.04998	.02422
0.31	.43512	.30705	.20018	.12017	.06624	.03346
0.32	.48895	.35655	.24085	.15010	.08604	.04525
0.33	.54238	.40801	.28510	.18420	.10965	.05998
0.34	.59453	.46055	.33236	.22230	.13726	.07799
0.35	.64459	.51327	.38196	.26411	.16892	.09958
0.36	.69187	.56531	.43310	.30916	.20457	.12499
0.37	.73584	.61584	.48496	.35685	.24400	.15432
0.38	.77609	.66413	.53669	.40652	.28684	.18760
0.39	.81239	.70956	.58747	.45737	.33261	.22469
0.40	.84463	.75165	.63653	.50862	.38070	.26533
0.41	.87285	.79005	.68320	.55943	.43043	.30914
0.42	.89720	.82458	.72692	.60905	.48102	.35559
0.43	.91790	.85516	.76727	.65676	.53170	.40407
0.44	.93524	.88185	.80394	.70192	.58169	.45387
0.45	.94956	.90480	.83679	.74405	.63023	.50424
0.46	.96122	.92426	.86578	.78276	.67667	.55441
0.47	.94050	.97957	.89098	.81779	.72043	.60362
0.48	.97797	.95387	.91257	.84903	.76104	.65116
0.49	.98373	.96472	.93079	.87646	.79817	.69640
0.50	.98815	.97338	.94594	.90021	.83161	.73880

(*table continues*)

Table 5-4. Binomial Loss Probability (S = 40 Sources *Continued*)

Traffic per Source (a in Erl)	Loss Probability (P) for $N =$					
	19	20	21	22	23	24
0.11						
0.12						
0.13						
0.14						
0.15						
0.16						
0.17						
0.18	.00001					
0.19	.00003	.00001				
0.20	.00006	.00001				
0.21	.00011	.00003	.00001			
0.22	.00021	.00006	.00001			
0.23	.00039	.00011	.00003	.00001		
0.24	.00069	.00021	.00006	.00001		
0.25	.00117	.00037	.00011	.00003	.00001	
0.26	.00193	.00065	.00020	.00005	.00001	
0.27	.00309	.00109	.00035	.00010	.00003	.00001
0.28	.00480	.00177	.00060	.00018	.00005	.00001
0.29	.00727	.00281	.00099	.00032	.00009	.00002
0.30	.01073	.00434	.00160	.00054	.00016	.00004
0.31	.01547	.00653	.00251	.00088	.00028	.00008
0.32	.02180	.00960	.00385	.00141	.00047	.00014
0.33	.03007	.01380	.00578	.00220	.00076	.00024
0.34	.04067	.01942	.00847	.00337	.00122	.00040
0.35	.05395	.02679	.01216	.00504	.00190	.00065
0.36	.07027	.03625	.01712	.00738	.00289	.00103
0.37	.08995	.04817	.02363	.01059	.00432	.00160
0.38	.11326	.06290	.03203	.01491	.00633	.00244
0.39	.14036	.08076	.04266	.02062	.00910	.00365
0.40	.17133	.10206	.05588	.02803	.01283	.00535
0.41	.20613	.12700	.07202	.03745	.01779	.00770
0.42	.24458	.15573	.09139	.04923	.02426	.01089
0.43	.28639	.18827	.11425	.06373	.03254	.01515
0.44	.33113	.22454	.14079	.08125	.04297	.02075
0.45	.37825	.26431	.17110	.10209	.05589	.02796
0.46	.42711	.30726	.20517	.12648	.07164	.03711
0.47	.47701	.35292	.24288	.15458	.09052	.04853
0.48	.52719	.40071	.28396	.18646	.11282	.06257
0.49	.57689	.44998	.32805	.22205	.13873	.07956
0.50	.62537	.50000	.37463	.26120	.16839	.09980

Table 5-5. Binomial Loss Probability (S = 50 Sources)

Traffic per Source (a in Erl)	Loss Probability (P) for $N =$					
	15	16	17	18	19	20
0.11	.00017	.00004	.00001			
0.12	.00044	.00012	.00004	.00001		
0.13	.00102	.00031	.00009	.00002	.00001	
0.14	.00218	.00072	.00022	.00006	.00002	
0.15	.00429	.00154	.00051	.00015	.00004	.00001
0.16	.00787	.00303	.00107	.00035	.00011	.00003
0.17	.01360	.00561	.00213	.00074	.00024	.00007
0.18	.02226	.00978	.00396	.00148	.00051	.00016
0.19	.03469	.01619	.00696	.00276	.00101	.00034
0.20	.05175	.02557	.01166	.00491	.00191	.00069
0.21	.07419	.03870	.01866	.00831	.00343	.00131
0.22	.10260	.05636	.02865	.01347	.00586	.00236
0.23	.13730	.07922	.04235	.02097	.00962	.00409
0.24	.17830	.10779	.06047	.03146	.01518	.00679
0.25	.22527	.14236	.08362	.04562	.02310	.01086
0.26	.27752	.18291	.11227	.06410	.03401	.01676
0.27	.33405	.22913	.14666	.08746	.04853	.02503
0.28	.39366	.28039	.18678	.11612	.06726	.03627
0.29	.45494	.33576	.23232	.15030	.09075	.05106
0.30	.51646	.39412	.28269	.19000	.11937	.06999
0.31	.57684	.45417	.33705	.23492	.15334	.09354
0.32	.63478	.51456	.39433	.28451	.19263	.12208
0.33	.68922	.57396	.45332	.33798	.23698	.15582
0.34	.73933	.63116	.51274	.39432	.28588	.19472
0.35	.78452	.68509	.57132	.45241	.33857	.23856
0.36	.82449	.73494	.62789	.51100	.39412	.28684
0.37	.85917	.78009	.68140	.56889	.45142	.33886
0.38	.88869	.82021	.73104	.62493	.50933	.39372
0.39	.91335	.85519	.77617	.67811	.56664	.45037
0.40	.93358	.88511	.81644	.72758	.62226	.50770
0.41	.94988	.91023	.85169	.77272	.67517	.56455
0.42	.96277	.93094	.88198	.81314	.72453	.61983
0.43	.97278	.94771	.90751	.84865	.76971	.67255
0.44	.98042	.96103	.92865	.87927	.81029	.72186
0.45	.98615	.97142	.94583	.90518	.84605	.76711
0.46	.99036	.97939	.95953	.92670	.87697	.80786
0.47	.99340	.98538	.97027	.94424	.90322	.84386
0.48	.99557	.98981	.97852	.95828	.92508	.87507
0.49	.99707	.99302	.98474	.96931	.94295	.90163
0.50	.99810	.99530	.98935	.97781	.95728	.92380

(*table continues*)

Table 5-5. Binomial Loss Probability (S = 50 Sources *Continued*)

Traffic per Source (*a* in Erl)	Loss Probability (*P*) for *N* =					
	21	22	23	24	24	26
0.11						
0.12						
0.13						
0.14						
0.15						
0.16	.00001					
0.17	.00002					
0.18	.00005	.00001				
0.19	.00011	.00003	.00001			
0.20	.00023	.00007	.00002	.00001		
0.21	.00046	.00015	.00005	.00001		
0.22	.00080	.00030	.00010	.00003	.00001	
0.23	.00161	.00059	.00020	.00006	.00002	
0.24	.00282	.00108	.00039	.00013	.00004	.00001
0.25	.00473	.00191	.00072	.00025	.00008	.00002
0.26	.00767	.00325	.00125	.00047	.00016	.00005
0.27	.01200	.00534	.00221	.00085	.00030	.00010
0.28	.01818	.00847	.00367	.00147	.00055	.00019
0.29	.02675	.01303	.00590	.00248	.00097	.00035
0.30	.03824	.01945	.00920	.00405	.00165	.00063
0.31	.05324	.02824	.01394	.00640	.00273	.00108
0.32	.07229	.03993	.02054	.00984	.00438	.00181
0.33	.09585	.05507	.02950	.01472	.00683	.00294
0.34	.12429	.07418	.04133	.02146	.01037	.00466
0.35	.15778	.09772	.05655	.03053	.01536	.00718
0.36	.19633	.12602	.07568	.04245	.02219	.01080
0.37	.23970	.15927	.09916	.05771	.03134	.01585
0.38	.28743	.19747	.12730	.07681	.04329	.02274
0.39	.33887	.24042	.16031	.10018	.05854	.03192
0.40	.39314	.28767	.19819	.12815	.07758	.04386
0.41	.44925	.33861	.24075	.16092	.10082	.05905
0.42	.50611	.39239	.28759	.19849	.12860	.07798
0.43	.56260	.44806	.33809	.24070	.16111	.10107
0.44	.61764	.50456	.39148	.28718	.19840	.12864
0.45	.67024	.56078	.44680	.33733	.24030	.16090
0.46	.71954	.61566	.50303	.39040	.28645	.19791
0.47	.76490	.66820	.55907	.44546	.33632	.23953
0.48	.80583	.71757	.61387	.50151	.38915	.28542
0.49	.84208	.76306	.66644	.55747	.44404	.33506
0.50	.87356	.80420	.71591	.61227	.50000	.38772

(*table continues*)

Table 5-5. Binomial Loss Probability (S = 50 Sources *Continued*)

Traffic per Source (a in Erl)	Loss Probability (P) for N =					
	27	28	29	30	31	32
0.11						
0.12						
0.13						
0.14						
0.15						
0.16						
0.17						
0.18						
0.19						
0.20						
0.21						
0.22						
0.23						
0.24						
0.25	.00001					
0.26	.00001					
0.27	.00003	.00001				
0.28	.00006	.00002				
0.29	.00012	.00004	.00001			
0.30	.00022	.00007	.00002	.00001		
0.31	.00040	.00013	.00004	.00001		
0.32	.00069	.00025	.00008	.00002	.00001	
0.33	.00118	.00044	.00015	.00005	.00001	
0.34	.00194	.00075	.00027	.00009	.00003	.00001
0.35	.00312	.00126	.00047	.00016	.00005	.00002
0.36	.00488	.00205	.00080	.00029	.00009	.00003
0.37	.00746	.00326	.00132	.00049	.00017	.00005
0.38	.01112	.00505	.00213	.00083	.00030	.00010
0.39	.01620	.00764	.00335	.00136	.00051	.00018
0.40	.02311	.01132	.00515	.00217	.00084	.00030
0.41	.03226	.01641	.00775	.00339	.00137	.00051
0.42	.04415	.02328	.01141	.00518	.00218	.00084
0.43	.05926	.03239	.01646	.00776	.00339	.00136
0.44	.07805	.04418	.02328	.01139	.00516	.00216
0.45	.10094	.05915	.03229	.01637	.00769	.00334
0.46	.12829	.07777	.04395	.02309	.01125	.00506
0.47	.16030	.10045	.05875	.03197	.01614	.00754
0.48	.19705	.12755	.07715	.04346	.02273	.01100
0.49	.23841	.15931	.09959	.05804	.03144	.01577
0.50	.28409	.19580	.12643	.07620	.04272	.02219

Table 5-6. Binomial Loss Probability (S = 60 Sources)

Traffic per Source (a in Erl)	Loss Probability (P) for N =					
	16	18	20	22	24	26
0.11	.00048	.00004				
0.12	.00125	.00013	.00001			
0.13	.00287	.00035	.00003			
0.14	.00601	.00086	.00009	.00001		
0.15	.01154	.00193	.00024	.00002		
0.16	.02060	.00396	.00057	.00006	.00001	
0.17	.03443	.00757	.00125	.00016	.00002	
0.18	.05431	.01354	.00225	.00036	.00004	
0.19	.08137	.02283	.00485	.00079	.00010	.00001
0.20	.11639	.03651	.00872	.00159	.00022	.00002
0.21	.15967	.05567	.01485	.00304	.00048	.00006
0.22	.21096	.08126	.02409	.00549	.00096	.00013
0.23	.26940	.11401	.03736	.00945	.00185	.00028
0.24	.33356	.15426	.05561	.01555	.00337	.00057
0.25	.40165	.20189	.07970	.02453	.00587	.00109
0.26	.47157	.25631	.11032	.03721	.00979	.00201
0.27	.54122	.31644	.14784	.05445	.01571	.00354
0.28	.60858	.38079	.19227	.07705	.02428	.00599
0.29	.67195	.44761	.24320	.10564	.03627	.00978
0.30	.72996	.51502	.29982	.14066	.05244	.01541
0.31	.78173	.58117	.36092	.18221	.07353	.02349
0.32	.82678	.64437	.42499	.23004	.10020	.03469
0.33	.86505	.70321	.49038	.28355	.13288	.04974
0.34	.89679	.75664	.55536	.34175	.17179	.06936
0.35	.92253	.80398	.61830	.40335	.21681	.09416
0.36	.94292	.84494	.67775	.46688	.26749	.12464
0.37	.95873	.87957	.73255	.53075	.32305	.16108
0.38	.97072	.90817	.78188	.59337	.38240	.20348
0.39	.97962	.93128	.82525	.65330	.44422	.25155
0.40	.98608	.94953	.86250	.86250	.50703	.30467
0.41	.99068	.96364	.89379	.76039	.56932	.36191
0.42	.99388	.97430	.91948	.80595	.62964	.42212
0.43	.99606	.98219	.94010	.84565	.68669	.48392
0.44	.99751	.98790	.95629	.87947	.73941	.54588
0.45	.98846	.99194	.96873	.90753	.78702	.60655
0.46	.99907	.99474	.97806	.93055	.82903	.66458
0.47	.99945	.99664	.98492	.94880	.86526	.71883
0.48	.99968	.99789	.98985	.96300	.89581	.76838
0.49	.99982	.99871	.99331	.97380	.92099	.81262
0.50	.99990	.99923	.99568	.98183	.94126	.85120

(*table continues*)

Table 5-6. Binomial Loss Probability (S = 60 Sources *Continued*)

Traffic per Source (a in Erl)	Loss Probability (P) for N =					
	28	30	32	34	36	38
0.11						
0.12						
0.13						
0.14						
0.15						
0.16						
0.17						
0.18						
0.19						
0.20						
0.21	.00001					
0.22	.00001					
0.23	.00003					
0.24	.00007	.00001				
0.25	.00016	.00002				
0.26	.00032	.00004				
0.27	.00062	.00008	.00001			
0.28	.00115	.00017	.00002			
0.29	.00206	.00034	.00004			
0.30	.00355	.00064	.00009	.00001		
0.31	.00590	.00116	.00018	.00002		
0.32	.00947	.00203	.00034	.00004		
0.33	.01473	.00343	.00062	.00009	.00001	
0.34	.02224	.00562	.00111	.00017	.00002	
0.35	.03261	.00892	.00191	.00032	.00004	
0.36	.04652	.01376	.00320	.00058	.00008	.00001
0.37	.06466	.02063	.00518	.00102	.00015	.00002
0.38	.08765	.03013	.00818	.00173	.00028	.00004
0.39	.11601	.04288	.01255	.00288	.00051	.00007
0.40	.15010	.05955	.01877	.00464	.00089	.00013
0.41	.19003	.08077	.02736	.00730	.00151	.00024
0.42	.23563	.10708	.03894	.01118	.00250	.00043
0.43	.28644	.13890	.05414	.01671	.00403	.00075
0.44	.34170	.17644	.07360	.02439	.00633	.00127
0.45	.40037	.21967	.09789	.03479	.00971	.00210
0.46	.46121	.26826	.12749	.04853	.01455	.00338
0.47	.52284	.32159	.16270	.06623	.02130	.00532
0.48	.58383	.37878	.20360	.08852	.03052	.00820
0.49	.64281	.43869	.25001	.11591	.04278	.01234
0.50	.69854	.50000	.30146	.14880	.05874	.01817

6

Engset Distribution

The Engset distribution, also known as the Erlang-Engset distribution, is used in lieu of the Erlang B distribution for dimensioning nonqueued (immediate service) common equipment pools that have a limited number of sources, such as small rural switching units and line concentrators. It is based on the following assumptions:

- Calls are served in random order.
- There are a finite number of sources.
- Blocked calls are cleared.
- Holding time is constant or exponential.

6.1   ENGSET FORMULA

The Engset formula, also known as the *truncated Binomial formula*, is given in Equation 6.1. Note that P is present on both sides of the equation, requiring an iterative (i.e., successive approximation) calculation process. When P is very small, such that the value of $1 - P$ approaches unity, the approximate Engset formula of Equation 6.2 can be used.

$$P = \frac{\dfrac{(S-1)!}{N!\,(S-1-N)!}\left[\dfrac{A}{S-A\,(1-P)}\right]^N}{\displaystyle\sum_{i=0}^{N}\dfrac{(S-1)!}{i!\,(S-1-i)!}\left[\dfrac{A}{S-A\,(1-P)}\right]^i} \qquad (6.1)$$

$$P \approx \frac{\dfrac{(S-1)!}{N!\,(S-1-N)!}\left[\dfrac{A}{S-A}\right]^N}{\displaystyle\sum_{i=0}^{N}\dfrac{(S-1)!}{i!\,(S-1-i)!}\left[\dfrac{A}{S-A}\right]^i} \qquad (6.2)$$

where P = Engset loss probability
$\quad$ S = Number of sources
$\quad$ N = Number of servers in full-availability group
$\quad$ A = Traffic offered to group in Erlangs

6.2 ENGSET COMPUTER PROGRAM

The following iterative computer program can be used to calculate Equation 6.1 to determine Engset loss probabilities with an accuracy of ±0.00001 as set in Step 220. More accurate results can be obtained by reducing this parameter; this will result in more iterations and longer run times, however. A noniterative, faster program can be written to calculate Equation 6.2 but the results, even when P is very small, will be less accurate. Required inputs for this program are the number of sources, the number of servers in the group, and the traffic offered to the group expressed in Erlangs.

```
100 REM ENGSET LOSS PROBABILITY CALCULATION
110 INPUT "ENTER NUMBER OF SOURCES (S)";S
120 INPUT "ENTER NUMBER OF SERVERS (N)";N
130 INPUT "ENTER OFFERED TRAFFIC IN ERLANGS (A)";A
140 LET P0=P1
150 LET X=1
160 LET Y=1
170 FOR I=1 TO N
180 LET X=X*A/((S-A*(1-P0))*(S-I)/I
```

```
190 LET Y=X+Y
200 NEXT I
210 LET P1=X/Y
220 IF ABS(P0-P1)>.00001 GOTO 140
230 PRINT USING "P = #.#####";P1
240 END
```

6.3   ENGSET LOSS PROBABILITY TABLES

Engset loss probability tables (Tables 6-1 through 6-6) are used to determine the loss probability (grade of service) when traffic is offered to a group of servers from a finite group of sources. Blanks (no data) in the tables indicate that the loss probability is less than 0.000005—not necessarily zero, but essentially nonblocking for practical applications. The following examples illustrate typical table usage:

Example 6-1

Determine the loss probability for a line concentrator with 60 subscribers, 18 channels, and offered traffic of 0.2 Erlangs per subscriber (source).

$$A = (60 \text{ sub})(0.2 \text{ Erl/sub}) = 12 \text{ Erlangs}$$

Using Table 6-6 (60 sources) (pages 161–164), select the A row for 12 Erlangs and the P column for 18 servers, and read .01787 at the intersection.

Example 6-2

For the line concentrator of Example 6-1, determine the offered traffic in Erlangs and CCS to achieve a grade of service of 0.005 or better.

Using Table 6-6 (60 sources), select the N column for 18 servers and read down until .00436 is found. Read across the A row to determine that 10.2 Erlangs of traffic can be offered.

$$(10.2 \text{ Erl})(36 \text{ CCS/Erl}) = 367.2 \text{ CCS}$$

Example 6-3

Determine the number of trunks required for a 40-line remote switching unit if nonlocal busy-hour traffic is 0.1 Erlangs per subscriber and a grade of service of 0.01 or better is required.

$$A = (40 \text{ lines})(0.1 \text{ Erl/line}) = 4 \text{ Erlangs}$$

Using Table 6-4 (40 sources) (pages 155–157), select the A row for 4 Erlangs and read over until .00897 is found. Read up the N column to determine that at least 9 trunks are required.

Table 6-1. Engset Loss Probability (S = 10 Sources)

Offered Traffic (A in Erl)	Loss Probability (P) for N =					
	2	3	4	5	6	7
0.1	.00336	.00008				
0.2	.01250	.00060	.00002			
0.3	.02619	.00189	.00009			
0.4	.04335	.00421	.00026	.00001		
0.5	.06305	.00772	.00061	.00003		
0.6	.08453	.01251	.00120	.00008		
0.7	.10717	.01862	.00210	.00016	.00001	
0.8	.13046	.02604	.00340	.00030	.00002	
0.0	.15401	.03471	.00515	.00051	.00003	
1.0	.17749	.04455	.00743	.00083	.00006	
1.1	.20070	.05544	.01028	.00127	.00010	.00001
1.2	.22344	.06727	.01374	.00188	.00017	.00001
1.3	.24561	.07991	.01785	.00268	.00027	.00002
1.4	.26711	.09323	.02263	.00370	.00040	.00003
1.5	.28789	.10709	.02809	.00498	.00059	.00004
1.6	.30792	.12139	.03421	.00655	.00083	.00007
1.7	.32720	.13599	.04099	.00845	.00116	.00010
1.8	.34571	.15080	.04840	.01070	.00157	.00015
1.9	.36347	.16573	.05640	.01332	.00210	.00021
2.0	.38050	.18069	.06495	.01635	.00275	.00029
2.1	.39681	.19562	.07401	.01980	.00354	.00040
2.2	.41244	.21044	.08353	.02367	.00450	.00055
2.3	.42740	.22512	.09344	.02799	.00564	.00073
2.4	.44174	.23961	.10371	.03276	.00699	.00095
2.5	.45547	.25387	.11427	.03797	.00857	.00123
2.6	.46862	.26788	.12507	.04361	.01039	.00158
2.7	.48123	.28161	.13607	.04967	.01248	.00200
2.8	.49332	.29505	.14721	.05614	.01485	.00251
2.9	.50491	.30819	.15846	.06300	.01751	.00311
3.0	.51603	.32103	.16976	.07021	.02049	.00383
3.1	.52671	.33351	.18109	.07776	.02379	.00468
3.2	.53696	.34572	.19241	.08562	.02741	.00566
3.3	.54682	.35760	.20369	.09375	.03137	.00680
3.4	.55629	.36916	.21491	.10213	.03566	.00811
3.5	.56540	.38041	.22604	.11072	.04029	.00961
3.6	.57417	.39135	.23707	.11949	.04524	.01130
3.7	.58261	.40199	.24797	.12842	.05051	.01321
3.8	.59075	.41234	.25874	.13747	.05609	.01535
3.9	.59859	.42239	.26935	.14662	.06196	.01772
4.0	.60615	.43216	.27980	.15584	.06812	.02035

Table 6-2. Engset Loss Probability (S = 20 Sources)

Offered Traffic (A in Erl)	Loss Probability (P) for N =					
	2	3	4	5	6	7
0.2	.01442	.00083	.00003			
0.4	.04874	.00561	.00046	.00003		
0.6	.09301	.01609	.00199	.00018	.00001	
0.8	.14103	.03237	.00538	.00067	.00007	.00001
1.0	.18920	.05371	.01123	.00177	.00022	.00002
1.2	.23553	.07894	.01987	.00380	.00057	.00007
1.4	.27908	.10685	.03140	.00707	.00124	.00017
1.6	.31947	.13634	.04567	.01185	.00240	.00039
1.8	.35665	.16652	.06236	.01832	.00423	.00078
2.0	.39076	.19669	.08108	.02658	.00689	.00142
2.2	.42200	.22635	.10136	.03664	.01054	.00242
2.4	.45061	.25515	.12276	.04842	.01533	.00389
2.6	.47684	.28288	.14488	.06176	.02134	.00593
2.8	.50092	.30939	.16736	.07646	.02863	.00866
3.0	.52307	.33464	.18991	.09230	.03719	.01219
3.2	.54348	.35860	.21229	.10904	.04699	.01658
3.4	.56233	.38130	.23432	.12646	.05793	.02192
3.6	.57977	.40276	.25586	.14433	.06992	.02823
3.8	.59596	.42305	.27681	.16247	.08282	.03552
4.0	.61100	.44222	.29710	.18071	.09648	.04377
4.2	.62501	.46033	.31670	.19891	.11076	.05293
4.4	.63808	.47745	.33558	.21695	.12552	.06294
4.6	.65031	.49364	.35374	.23475	.14062	.07371
4.8	.66177	.50896	.37118	.25222	.15595	.08516
5.0	.67252	.52346	.38790	.26931	.17138	.09718
5.2	.68264	.53720	.40394	.28599	.18684	.10967
5.4	.69216	.55024	.41931	.30222	.20223	.12253
5.6	.70114	.56261	.43403	.31798	.21749	.13576
5.8	.70962	.57437	.44813	.33327	.23256	.14901
6.0	.71765	.58554	.46165	.34808	.24740	.16247
6.2	.72526	.59618	.47460	.36241	.26198	.17597
6.4	.73247	.60632	.48701	.37627	.27626	.18946
6.6	.73932	.61598	.49891	.38967	.29022	.20288
6.8	.74583	.62520	.51032	.40261	.30385	.21619
7.0	.75204	.63401	.52127	.41511	.31715	.22935
7.2	.75795	.64243	.53179	.42719	.33010	.24234
7.4	.76359	.65049	.54189	.43884	.34270	.25513
7.6	.76898	.65821	.55159	.45010	.35496	.26769
7.8	.77413	.66560	.56092	.46098	.36688	.28001
8.0	.77906	.67269	.56990	.47148	.37845	.29209

(*table continues*)

Table 6-2. Engset Loss Probability (S = 20 Sources *Continued*)

Offered Traffic (A in Erl)	Loss Probability (P) for N =					
	8	9	10	11	12	13
0.2						
0.4						
0.6						
0.8						
1.0						
1.2	.00001					
1.4	.00002					
1.6	.00005	.00001				
1.8	.00012	.00001				
2.0	.00024	.00003				
2.2	.00045	.00007	.00001			
2.4	.00080	.00013	.00002			
2.6	.00133	.00024	.00004			
2.8	.00212	.00042	.00007	.00001		
3.0	.00323	.00070	.00012	.00002		
3.2	.00475	.00111	.00021	.00003		
3.4	.00676	.00170	.00035	.00006	.00001	
3.6	.00933	.00251	.00055	.00010	.00001	
3.8	.01254	.00361	.00085	.00016	.00003	
4.0	.01644	.00505	.00127	.00026	.00004	.00001
4.2	.02110	.00690	.00184	.00040	.00007	.00001
4.4	.02653	.00921	.00261	.00060	.00011	.00002
4.6	.03276	.01205	.00362	.00089	.00018	.00003
4.8	.03976	.01545	.00492	.00128	.00027	.00005
5.0	.04753	.01947	.00655	.00180	.00040	.00007
5.2	.05602	.02414	.00857	.00248	.00058	.00011
5.4	.06517	.02946	.01102	.00337	.00083	.00017
5.6	.07493	.03544	.01395	.00449	.00117	.00025
5.8	.08522	.04208	.01739	.00589	.00162	.00036
6.0	.09596	.04935	.02138	.00761	.00219	.00051
6.2	.10709	.05720	.02592	.00969	.00293	.00071
6.4	.11853	.06561	.03104	.01216	.00387	.00099
6.6	.13021	.07451	.03673	.01507	.00503	.00135
6.8	.14207	.08386	.04297	.01845	.00645	.00181
7.0	.15404	.09359	.04975	.02231	.00818	.00240
7.2	.16607	.10365	.05704	.02667	.01023	.00315
7.4	.17813	.11398	.06480	.03153	.01266	.00408
7.6	.19015	.12452	.07299	.03690	.01548	.00522
7.8	.20211	.13523	.08156	.04276	.01872	.00661
8.0	.21398	.14605	.09046	.04909	.02240	.00827

Table 6-3. Engset Loss Probability ($S = 30$ Sources)

Offered Traffic (A in Erl)	Loss Probability (P) for $N =$					
	4	5	6	7	8	9
0.3	.00018	.00001				
0.6	.00229	.00023	.00002			
0.9	.00899	.00139	.00017	.00002		
1.2	.02198	.00457	.00076	.00010	.00001	
1.5	.04151	.01084	.00228	.00039	.00006	.00001
1.8	.06672	.02096	.00534	.00112	.00020	.00003
2.1	.09619	.03518	.01053	.00260	.00054	.00009
2.4	.12837	.05327	.01831	.00523	.00125	.00025
2.7	.16189	.07466	.02893	.00937	.00255	.00059
3.0	.19567	.09861	.04239	.01536	.00469	.00122
3.3	.22893	.12434	.05847	.02342	.00795	.00229
3.6	.26115	.15111	.07682	.03365	.01257	.00400
3.9	.29201	.17830	.09697	.04599	.01875	.00654
4.2	.32133	.20543	.11846	.06028	.02661	.01010
4.5	.34903	.23213	.14082	.07628	.03619	.01485
4.8	.37513	.25813	.16366	.09367	.04744	.02093
5.1	.39964	.28327	.18663	.11214	.06023	.02841
5.4	.42265	.30742	.20947	.13138	.07438	.03730
5.7	.44424	.33053	.23196	.15110	.08967	.04756
6.0	.46448	.35259	.25395	.17105	.10588	.05909
6.3	.48348	.37559	.27532	.19102	.12277	.07176
6.6	.50132	.39356	.29601	.21084	.14013	.08540
6.9	.51809	.41252	.31597	.23039	.15778	.09984
7.2	.53386	.43054	.33518	.24954	.17553	.11490
7.5	.54871	.44764	.35362	.26825	.19326	.13042
7.8	.56270	.46387	.37131	.28643	.21085	.14624
8.1	.57592	.47929	.38826	.30408	.22820	.16223
8.4	.58840	.49395	.40449	.32115	.24525	.17826
8.7	.60021	.50788	.42002	.33765	.26194	.19425
9.0	.61139	.52113	.43489	.35357	.27823	.21011
9.3	.62200	.53374	.44911	.36891	.29409	.22576
9.6	.63206	.54576	.46273	.38369	.30950	.24117
9.9	.64163	.55721	.47577	.39793	.32446	.25628
10.2	.65072	.56814	.48825	.41163	.33896	.27107
10.5	.65939	.57857	.50021	.42482	.35300	.28551
10.8	.66764	.58854	.51167	.43751	.36659	.29960
11.1	.67552	.59807	.52266	.44972	.37974	.31331
11.4	.68305	.60719	.53321	.46148	.39245	.32665
11.7	.69024	.61592	.54333	.47280	.40473	.33961
12.0	.68712	.62429	.55305	.48370	.41661	.35220

(*table continues*)

Table 6-3. Engset Loss Probability (*S* = 30 Sources *Continued*)

Offered Traffic (*A* in Erl)	Loss Probability (*P*) for *N* =					
	10	11	12	13	14	15
0.3						
0.6						
0.9						
1.2						
1.5						
1.8						
2.1	.00001					
2.4	.00004	.00001				
2.7	.00012	.00002				
3.0	.00027	.00005	.00001			
3.3	.00057	.00012	.00002			
3.6	.00109	.00026	.00005	.00001		
3.9	.00196	.00051	.00011	.00002		
4.2	.00329	.00093	.00023	.00005	.00001	
4.5	.00525	.00160	.00042	.00010	.00002	
4.8	.00798	.00263	.00075	.00019	.00004	.00001
5.1	.01164	.00413	.00127	.00034	.00008	.00002
5.4	.01634	.00622	.00205	.00059	.00015	.00003
5.7	.02218	.00901	.00318	.00098	.00026	.00006
6.0	.02922	.01264	.00476	.00156	.00045	.00011
6.3	.03745	.01719	.00689	.00240	.00073	.00019
6.6	.04684	.02274	.00967	.00358	.00116	.00033
6.9	.05731	.02932	.01319	.00519	.00178	.00053
7.2	.06876	.03694	.01754	.00729	.00265	.00084
7.5	.08106	.04556	.02277	.01000	.00383	.00128
7.8	.09407	.05513	.02891	.01338	.00541	.00191
8.1	.10766	.06555	.03596	.01749	.00746	.00278
8.4	.12168	.07674	.04391	.02240	.01005	.00394
8.7	.13603	.08858	.05269	.02812	.01326	.00547
9.0	.15057	.10095	.06225	.03466	.01713	.00743
9.3	.16521	.11375	.07249	.04199	.02171	.00988
9.6	.17987	.12686	.08334	.05009	.02704	.01289
9.9	.19446	.14020	.09469	.05889	.03310	.01652
10.2	.20893	.15367	.10645	.06833	.03989	.02079
10.5	.22323	.16719	.11853	.07833	.04738	.02573
10.8	.23732	.18071	.13084	.08880	.05552	.03136
11.1	.25117	.19417	.14331	.09967	.06425	.03765
11.4	.26474	.20751	.15587	.11086	.07351	.04459
11.7	.27803	.22070	.16846	.12230	.08322	.05214
12.0	.29102	.23370	.18103	.13390	.09333	.06025

Table 6-4. Engset Loss Probability ($S = 40$ Sources)

Offered Traffic (A in Erl)	Loss Probability (P) for $N =$					
	4	5	6	7	8	9
0.4	.00058	.00004				
0.8	.00649	.00093	.00011	.00001		
1.2	.02303	.00497	.00087	.00013	.00002	
1.6	.05115	.01475	.00348	.00068	.00011	.00002
2.0	.08837	.03168	.00939	.00233	.00049	.00009
2.4	.13016	.05562	.01982	.00595	.00152	.00033
2.8	.17592	.08524	.03531	.01243	.00374	.00097
3.2	.22062	.11872	.05565	.02245	.00778	.00233
3.6	.26366	.15428	.08007	.03628	.01424	.00485
4.0	.30425	.19048	.10752	.05376	.02353	.00897
4.4	.34205	.22625	.13688	.07442	.03582	.01514
4.8	.37699	.26088	.16720	.09757	.05103	.02363
5.2	.40915	.29396	.19767	.12248	.06882	.03459
5.6	.43869	.32525	.22771	.14846	.08875	.04794
6.0	.46582	.35468	.25692	.17489	.11029	.06349
6.4	.49075	.38225	.28505	.20131	.13292	.08089
6.8	.51369	.40802	.31192	.22735	.15618	.09978
7.2	.53483	.43209	.33748	.25275	.17966	.11975
7.6	.55435	.45455	.36171	.27734	.20306	.14044
8.0	.57240	.47552	.38461	.30100	.22613	.16150
8.4	.58913	.49511	.40624	.32367	.24869	.18267
8.8	.60467	.51343	.42665	.34533	.27061	.20373
9.2	.61913	.53057	.44591	.36599	.29181	.22450
9.6	.63262	.54664	.46408	.38565	.31224	.24484
10.0	.64522	.56173	.48122	.40435	.33187	.26467
10.4	.65702	.57590	.49742	.42213	.35071	.28392
10.8	.66809	.58923	.51272	.43903	.36875	.30256
11.2	.67848	.60179	.52719	.45510	.38602	.32055
11.6	.68826	.61365	.54090	.47038	.40253	.33788
12.0	.69748	.62484	.55388	.48491	.41832	.35457
12.4	.70618	.63544	.56620	.49874	.43341	.37061
12.8	.71440	.64547	.57789	.51191	.44784	.38603
13.2	.72219	.65498	.58900	.52446	.46163	.40083
13.6	.72957	.66402	.59957	.53643	.47483	.41505
14.0	.73658	.67260	.60963	.54785	.48746	.42871
14.4	.74324	.68077	.61922	.55875	.49954	.44182
14.8	.74958	.68855	.62837	.56917	.51112	.45441
15.2	.75561	.68597	.63711	.57914	.52221	.46650
15.6	.76137	.70306	.64545	.58867	.53284	.47812
16.0	.76686	.70983	.65344	.59781	.54304	.48930

(*table continues*)

Table 6-4. Engset Loss Probability (*S* = 40 Sources *Continued*)

Offered Traffic (*A* in Erl)	Loss Probability (*P*) for *N* =					
	10	11	12	13	14	15
0.4						
0.8						
1.2						
1.6						
2.0	.00001					
2.4	.00006	.00001				
2.8	.00022	.00004	.00001			
3.2	.00061	.00014	.00003	.00001		
3.6	.00144	.00038	.00009	.00002		
4.0	.00299	.00088	.00023	.00005	.00001	
4.4	.00561	.00183	.00053	.00014	.00003	.00001
4.8	.00964	.00347	.00110	.00031	.00008	.00002
5.2	.01541	.00607	.00212	.00066	.00018	.00005
5.6	.02313	.00991	.00377	.00128	.00039	.00010
6.0	.03292	.01525	.00628	.00231	.00076	.00022
6.4	.04475	.02225	.00989	.00392	.00139	.00044
6.8	.05847	.03103	.01479	.00630	.00240	.00082
7.2	.07385	.04158	.02114	.00965	.00394	.00145
7.6	.09059	.05375	.02904	.01493	.00617	.00242
8.0	.10839	.06748	.03848	.01989	.00926	.00387
8.4	.12694	.08245	.04942	.02701	.01335	.00594
8.8	.14598	.09844	.06170	.03551	.01856	.00876
9.2	.16525	.11521	.07517	.04534	.02499	.01248
9.6	.18456	.13251	.08963	.05642	.03266	.01720
10.0	.20375	.15015	.10486	.06861	.04154	.02301
10.4	.22267	.16793	.12068	.08174	.05057	.02994
10.8	.24124	.18571	.13688	.09565	.06264	.03798
11.2	.25939	.20335	.15332	.11016	.07462	.04709
11.6	.27706	.22077	.16984	.12511	.08737	.05718
12.0	.29421	.23789	.18634	.14035	.10073	.06815
12.4	.31083	.25464	.20270	.15575	.11457	.07987
12.8	.32691	.27099	.21886	.17120	.12874	.09221
13.2	.34244	.28691	.23476	.18660	.14313	.10505
13.6	.35743	.30237	.25034	.20188	.15763	.11826
14.0	.37190	.31738	.26557	.21697	.17215	.13173
14.4	.38584	.33192	.28044	.23183	.18661	.14536
14.8	.39928	.34600	.29491	.24641	.20095	.15907
15.2	.41222	.35962	.30900	.26069	.21512	.17278
15.6	.42470	.37280	.32268	.27465	.22908	.18642
16.0	.43673	.38554	.33596	.28827	.24280	.19995

(*table continues*)

Table 6-4. Engset Loss Probability ($S = 40$ Sources *Continued*)

Offered Traffic (A in Erl)	Loss Probability (P) for N =					
	16	17	18	19	20	21
0.4						
0.8						
1.2						
1.6						
2.0						
2.4						
2.8						
3.2						
3.6						
4.0						
4.4						
4.8						
5.2	.00001					
5.6	.00003	.00001				
6.0	.00006	.00001				
6.4	.00013	.00003	.00001			
6.8	.00026	.00007	.00002			
7.2	.00048	.00014	.00004	.00001		
7.6	.00085	.00027	.00008	.00002		
8.0	.00146	.00049	.00015	.00004	.00001	
8.4	.00238	.00086	.00028	.00008	.00002	.00001
8.8	.00372	.00143	.00049	.00015	.00004	.00001
9.2	.00562	.00228	.00084	.00028	.00008	.00002
9.6	.00819	.00352	.00136	.00048	.00015	.00004
10.0	.01156	.00525	.00215	.00079	.00026	.00008
10.4	.01583	.00758	.00327	.00128	.00045	.00014
10.8	.02108	.01063	.00484	.00199	.00074	.00025
11.2	.02735	.01448	.00694	.00300	.00117	.00041
11.6	.03465	.01922	.00968	.00441	.00181	.00067
12.0	.04292	.02490	.01316	.00629	.00272	.00106
12.4	.05216	.03153	.01744	.00875	.00397	.00162
12.8	.06222	.03908	.02257	.01187	.00565	.00242
13.2	.07303	.04752	.02859	.01573	.00784	.00353
13.6	.08445	.05677	.03548	.02037	.01064	.00502
14.0	.09639	.06674	.04321	.04321	.02583	.00697
14.4	.10873	.07734	.05172	.03211	.01828	.00945
14.8	.12137	.08847	.06095	.03920	.02323	.01255
15.2	.13420	.10001	.07079	.04703	.02896	.01631
15.6	.14716	.11188	.08117	.05557	.03544	.02079
16.0	.16017	.12399	.09198	.06472	.04266	.02599

Table 6-5. Engset Loss Probability (S = 50 Sources)

Offered Traffic (A in Erl)	Loss Probability (P) for N =					
	8	9	10	11	12	13
2.5	.00215	.00052	.00011	.00002		
3.0	.00600	.00174	.00044	.00010	.00002	
3.5	.01329	.00455	.00137	.00037	.00009	.00002
4.0	.02493	.00982	.00341	.00105	.00029	.00007
4.5	.04121	.01835	.00724	.00254	.00079	.00022
5.0	.06183	.03059	.01350	.00531	.00187	.00059
5.5	.08602	.04659	.02270	.00991	.00388	.00137
6.0	.11279	.06599	.03507	.01681	.00725	.00282
6.5	.14117	.08817	.05054	.02633	.01241	.00528
7.0	.17029	.11238	.06875	.03858	.01969	.00911
7.5	.19947	.13790	.08921	.05341	.02932	.01465
8.0	.22819	.16407	.11133	.07054	.04129	.02215
8.5	.25612	.19038	.13455	.08952	.05549	.03174
9.0	.28302	.21641	.15837	.10991	.07162	.04339
9.5	.30877	.24187	.18235	.13124	.08934	.05695
10.0	.33329	.26657	.20618	.15312	.10826	.07219
10.5	.35657	.29037	.22960	.17519	.12801	.08881
11.0	.37864	.31321	.25245	.19718	.14826	.10647
11.5	.39952	.33504	.27459	.21889	.16872	.12488
12.0	.41927	.35587	.29595	.24015	.18917	.14375
12.5	.43795	.37571	.31649	.26085	.20940	.16284
13.0	.45561	.39459	.33620	.28092	.22929	.18195
13.5	.47231	.41254	.35507	.30031	.24874	.20092
14.0	.48812	.42961	.37312	.31899	.26766	.21962
14.5	.50309	.44585	.39036	.33697	.28602	.23796
15.0	.51728	.46128	.40684	.35423	.30378	.30378
15.5	.53074	.47597	.42257	.37080	.32093	.27329
16.0	.54352	.48995	.43760	.38669	.33747	.29021
16.5	.55566	.50326	.45196	.40193	.35339	.30661
17.0	.56721	.51595	.46457	.41653	.36872	.32246
17.5	.57821	.52805	.47878	.43053	.38347	.33779
18.0	.58868	.39765	.53960	.44395	.49132	.35258
18.5	.59867	.55063	.50331	.45682	.41128	.36685
19.0	.60821	.56118	.51479	.46916	.42439	.38061
19.5	.61732	.57126	.52579	.48100	.43699	.39387
20.0	.62602	.58091	.53634	.49237	.44912	.40666
20.5	.63436	.59016	.54644	.50329	.46078	.41899
21.0	.64234	.59902	.55615	.51378	.47200	.43088
21.5	.64999	.60752	.56546	.52387	.48281	.44234
22.0	.65733	.61568	.57441	.53357	.49321	.45340

(table continues)

Table 6-5. Engset Loss Probability (S = 50 Sources *Continued*)

Offered Traffic (A in Erl)	Loss Probability (P) for N =					
	14	15	16	17	18	19
2.5						
3.0						
3.5						
4.0	.00002					
4.5	.00007	.00002				
5.0	.00017	.00004	.00001			
5.5	.00043	.00013	.00003	.00001		
6.0	.00099	.00031	.00009	.00002	.00001	
6.5	.00203	.00071	.00022	.00007	.00002	
7.0	.00382	.00145	.00050	.00016	.00005	.00001
7.5	.00665	.00274	.00103	.00035	.00011	.00003
8.0	.01083	.00482	.00196	.00072	.00025	.00008
8.5	.01663	.00795	.00347	.00138	.00050	.00017
9.0	.02422	.01239	.00579	.00247	.00097	.00035
9.5	.03368	.01834	.00915	.00418	.00175	.00067
10.0	.04497	.02594	.01377	.00670	.00299	.00122
10.5	.05795	.03523	.01981	.01024	.00486	.00211
11.0	.07238	.04615	.02736	.01497	.00754	.00348
11.5	.08803	.05857	.03645	.02104	.01120	.00548
12.0	.10461	.07229	.04700	.02851	.01601	.00829
12.5	.12186	.08708	.05890	.03738	.02207	.01205
13.0	.13955	.10271	.07197	.04758	.02942	.01689
13.5	.15745	.11896	.08600	.05900	.03807	.02291
14.0	.17541	.13561	.10080	.07147	.04793	.03013
14.5	.19326	.15248	.11616	.08483	.05890	.03855
15.0	.21091	.16942	.13189	.09888	.07084	.04809
15.5	.22827	.18630	.14786	.11346	.08359	.05865
16.0	.24527	.20302	.16390	.12839	.09697	.07010
16.5	.26186	.21950	.17992	.14354	.11085	.08229
17.0	.27802	.23568	.19581	.15879	.12506	.09508
17.5	.29371	.25152	.21150	.17403	.13950	.10833
18.0	.30894	.26697	.22695	.18918	.15403	.12190
18.5	.32369	.28203	.24209	.20417	.16857	.13568
19.0	.33797	.29667	.25691	.21894	.18305	.14957
19.5	.35179	.31089	.27137	.23346	.19739	.16349
20.0	.36514	.32469	.28547	.24769	.21156	.17736
20.5	.37805	.33807	.29920	.26160	.22550	.19112
21.0	.39052	.35103	.31254	.27520	.23919	.20472
21.5	.40257	.36359	.32550	.28845	.25260	.21813
22.0	.41421	.37574	.33809	.30137	.26573	.23132

(*table continues*)

Table 6-5. Engset Loss Probability (S = 50 Sources *Continued*)

Offered Traffic (A in Erl)	Loss Probability (P) for N =					
	20	21	22	23	24	25
2.5						
3.0						
3.5						
4.0						
4.5						
5.0						
5.5						
6.0						
6.5						
7.0						
7.5	.00001					
8.0	.00002	.00001				
8.5	.00005	.00001				
9.0	.00011	.00003	.00001			
9.5	.00024	.00008	.00002	.00001		
10.0	.00046	.00016	.00005	.00001		
10.5	.00084	.00031	.00010	.00003	.00001	
11.0	.00148	.00058	.00021	.00007	.00002	.00001
11.5	.00247	.00102	.00039	.00014	.00004	.00001
12.0	.00394	.00173	.00070	.00026	.00009	.00003
12.5	.00606	.00280	.00119	.00047	.00017	.00006
13.0	.00895	.00437	.00196	.00081	.00031	.00011
13.5	.01277	.00657	.00311	.00136	.00054	.00020
14.0	.01762	.00953	.00475	.00218	.00092	.00036
14.5	.02358	.01338	.00701	.00339	.00151	.00062
15.0	.03067	.01822	.01002	.00508	.00238	.00102
15.5	.03886	.02410	.01388	.00739	.00363	.00164
16.0	.04809	.03104	.01869	.01043	.00537	.00254
16.5	.05827	.03901	.02448	.01429	.00771	.00383
17.0	.06927	.04796	.03127	.01904	.01075	.00560
17.5	.08096	.05778	.03903	.02474	.01459	.00796
18.0	.09321	.06837	.04770	.03138	.01929	.01100
18.5	.10589	.07960	.05719	.03893	.02488	.01480
19.0	.11887	.09135	.06740	.04734	.03137	.01944
19.5	.13207	.10351	.07821	.05652	.03873	.02492
20.0	.14538	.11597	.08951	.06638	.04689	.03126
20.5	.15872	.12864	.10121	.07680	.05578	.03843
21.0	.17203	.14142	.11319	.08769	.06531	.04635
21.5	.18526	.15424	.12536	.09896	.07538	.05497
22.0	.19836	.16705	.13766	.11050	.08589	.06420

Table 6-6. Engset Loss Probability ($S = 60$ Sources)

Offered Traffic (A in Erl)	Loss Probability (P) for $N =$					
	9	10	11	12	13	14
3.6	.00574	.00183	.00052	.00013	.00003	.00001
4.2	.01350	.00507	.00170	.00051	.00014	.00003
4.8	.02630	.01136	.00439	.00153	.00048	.00014
5.4	.04449	.02167	.00950	.00376	.00134	.00044
6.0	.06761	.03648	.01785	.00791	.00318	.00116
6.6	.09460	.05564	.02997	.01470	.00656	.00266
7.2	.12421	.07852	.04591	.02464	.01209	.00541
7.8	.15523	.10420	.06534	.03794	.02026	.00992
8.4	.18669	.13169	.08760	.05444	.03136	.01665
9.0	.21784	.16012	.11191	.07372	.04538	.02591
9.6	.24817	.18877	.13752	.09518	.06206	.03780
10.2	.27737	.21710	.16377	.11819	.08096	.05221
10.8	.30523	.24470	.19010	.14215	.10158	.06883
11.4	.33168	.27144	.21613	.16653	.12337	.08724
12.0	.35670	.29705	.24156	.19093	.14586	.10701
12.6	.38031	.32149	.26619	.21503	.16864	.12768
13.2	.40255	.34473	.28991	.23861	.19139	.14888
13.8	.42349	.36679	.31265	.26150	.21386	.17028
14.4	.44320	.38770	.33437	.28361	.23586	.19160
15.0	.46177	.40749	.35508	.30488	.25727	.21266
15.6	.47926	.42622	.37480	.32529	.27800	.23330
16.2	.49575	.44395	.39356	.34482	.29800	.25342
16.8	.51131	.46074	.41140	.36349	.31724	.27295
17.4	.52601	.47665	.42835	.38131	.33573	.29183
18.0	.53991	.49172	.44448	.39833	.35345	.31006
18.6	.55306	.50602	.45981	.41457	.37044	.32761
19.2	.56552	.51959	.47440	.43006	.38671	.34449
19.8	.57734	.53248	.48829	.44485	.40228	.36071
20.4	.58856	.54475	.50152	.45896	.41718	.37629
21.0	.59922	.55641	.51413	.47245	.43145	.39124
21.6	.60937	.56753	.52616	.48533	.44511	.40560
22.2	.61903	.57812	.53764	.49764	.45819	.41837
22.8	.62825	.58824	.54861	.50942	.47073	.43260
23.4	.63704	.59789	.55910	.52070	.48274	.44530
24.0	.64543	.60712	.56913	.53150	.49427	.45749
24.6	.65346	.61579	.57874	.54184	.50532	.46921
25.2	.66114	.62441	.58794	.55177	.51593	.48047
25.8	.66849	.63251	.59676	.56129	.52612	.49130
26.4	.67554	.64027	.60523	.57044	.53592	.50171
27.0	.68230	.64773	.61336	.57922	.54534	.51174

(table continues)

Table 6-6. Engset Loss Probability ($S = 60$ Sources *Continued*)

Offered Traffic (A in Erl)	Loss Probability (P) for $N =$					
	15	16	17	18	19	20
3.6						
4.2	.00001					
4.8	.00004	.00001				
5.4	.00013	.00004	.00001			
6.0	.00039	.00012	.00003	.00001		
6.6	.00099	.00034	.00011	.00003	.00001	
7.2	.00222	.00083	.00029	.00009	.00003	.00001
7.8	.00445	.00183	.00069	.00024	.00008	.00002
8.4	.00812	.00364	.00150	.00057	.00020	.00007
9.0	.01365	.00663	.00296	.00122	.00047	.00016
9.6	.02138	.01117	.00539	.00240	.00099	.00038
10.2	.03146	.01760	.00912	.00436	.00193	.00079
10.8	.04387	.02613	.01446	.00741	.00352	.00155
11.4	.05844	.03680	.02164	.01183	.00600	.00282
12.0	.07484	.04953	.03079	.01787	.00965	.00484
12.6	.09271	.06409	.04188	.02569	.01471	.00784
13.2	.11165	.08019	.05476	.03531	.02135	.01205
13.8	.13131	.09749	.06922	.04666	.02966	.01768
14.4	.15136	.11566	.08496	.05960	.03963	.02483
15.0	.17154	.13439	.10170	.07387	.05115	.03354
15.6	.19162	.15341	.11914	.08923	.06405	.04375
16.2	.21146	.17251	.13702	.10541	.07809	.05535
16.8	.23091	.19151	.15512	.12217	.09306	.06815
17.4	.24991	.21026	.17326	.13927	.10871	.08193
18.0	.26837	.22868	.19129	.15655	.12483	.09650
18.6	.28627	.24668	.20909	.17383	.14123	.11164
19.2	.30358	.26420	.22658	.19099	.15775	.12717
19.8	.32030	.28122	.24369	.20795	.17425	.14292
20.4	.33641	.29771	.26038	.22461	.19064	.15875
21.0	.35194	.31367	.27661	.24092	.20683	.17456
21.6	.36688	.32909	.29236	.25685	.22274	.19024
22.2	.38127	.34398	.30764	.27237	.23834	.20574
22.8	.39511	.35835	.32243	.28746	.25359	.22098
23.4	.40842	.37220	.33673	.30211	.26846	.23593
24.0	.42124	.38556	.35056	.31632	.28294	.25055
24.6	.43356	.39845	.36393	.33008	.29702	.26483
25.2	.44543	.41086	.37684	.34342	.31069	.27875
25.8	.45685	.42284	.38931	.35632	.32396	.29231
26.4	.46785	.43438	.40135	.36881	.33684	.30549
27.0	.47845	.44552	.41299	.38090	.34932	.31830

(*table continues*)

Table 6-6. Engset Loss Probability (S = 60 Sources *Continued*)

Offered Traffic (A in Erl)	Loss Probability (P) for $N =$					
	21	22	23	24	25	26
3.6						
4.2						
4.8						
5.4						
6.0						
6.6						
7.2						
7.8	.00001					
8.4	.00002	.00001				
9.0	.00005	.00002				
9.6	.00013	.00004	.00001			
10.2	.00030	.00011	.00004	.00001		
10.8	.00063	.00024	.00008	.00003	.00001	
11.4	.00123	.00050	.00019	.00007	.00002	.00001
12.0	.00225	.00097	.00039	.00015	.00005	.00002
12.6	.00388	.00179	.00077	.00031	.00011	.00004
13.2	.00634	.00310	.00141	.00060	.00024	.00009
13.8	.00983	.00509	.00246	.00110	.00046	.00018
14.4	.01457	.00798	.00407	.00193	.00086	.00035
15.0	.02068	.01194	.00644	.00323	.00151	.00066
15.6	.02826	.01715	.00974	.00516	.00255	.00118
16.2	.03727	.02369	.01414	.00789	.00411	.00200
16.8	.04766	.03161	.01977	.01159	.00635	.00325
17.4	.05926	.04086	.02668	.01639	.00944	.00508
18.0	.07192	.05134	.03487	.02239	.01351	.00763
18.6	.08543	.06291	.04429	.02960	.01867	.01105
19.2	.09961	.07540	.05481	.03802	.02499	.01547
19.8	.11426	.08862	.06630	.04755	.03246	.02096
20.4	.12924	.10241	.07860	.05807	.04104	.02755
21.0	.14439	.11661	.09154	.06946	.05064	.03523
21.6	.15960	.13108	.10496	.08155	.06114	.04394
22.2	.17477	.14568	.11873	.09421	.07240	.05357
22.8	.18981	.16031	.13271	.10727	.08428	.06401
23.4	.20468	.17489	.14680	.12063	.09665	.07514
24.0	.21930	.18936	.16091	.13417	.10938	.08681
24.6	.23366	.20364	.17496	.14779	.12235	.09890
25.2	.24772	.21771	.18888	.16140	.13547	.11130
25.8	.26146	.23153	.20264	.17496	.14865	.12391
26.4	.27487	.24506	.21619	.18839	.16181	.13663
27.0	.28794	.25831	.22951	.20167	.17491	.14940

(table continues)

Table 6-6. Engset Loss Probability (S = 60 Sources *Continued*)

Offered Traffic (A in Erl)	Loss Probability (P) for N =					
	27	28	29	30	31	32
3.6						
4.2						
4.8						
5.4						
6.0						
6.6						
7.2						
7.8						
8.4						
9.0						
9.6						
10.2						
10.8						
11.4						
12.0	.00001					
12.6	.00001					
13.2	.00003	.00001				
13.8	.00007	.00002	.00001			
14.4	.00014	.00005	.00002	.00001		
15.0	.00027	.00010	.00004	.00001		
15.6	.00051	.00020	.00008	.00003	.00001	
16.2	.00091	.00038	.00015	.00006	.00002	.00001
16.8	.00155	.00069	.00029	.00011	.00004	.00001
17.4	.00255	.00119	.00052	.00021	.00008	.00003
18.0	.00403	.00198	.00091	.00039	.00016	.00006
18.6	.00612	.00317	.00153	.00069	.00029	.00011
19.2	.00898	.00487	.00247	.00117	.00052	.00021
19.8	.01273	.00723	.00384	.00190	.00088	.00038
20.4	.01746	.01039	.00578	.00300	.00145	.00066
21.0	.02323	.01443	.00840	.00457	.00232	.00110
21.6	.03006	.01945	.01183	.00674	.00358	.00178
22.2	.03791	.02548	.01616	.00962	.00535	.00278
22.8	.04671	.03252	.02144	.01331	.00774	.00420
23.4	.05635	.04050	.02770	.01790	.01087	.00617
24.0	.06671	.04935	.03489	.02342	.01481	.00878
24.6	.07768	.05897	.04298	.02986	.01963	.01214
25.2	.08914	.06924	.05186	.03720	.02536	.01632
25.8	.10096	.08006	.06144	.04536	.03197	.02136
26.4	.11305	.09130	.07162	.05425	.03942	.02726
27.0	.12531	.10286	.08227	.06378	.04763	.03402

Definitions

Address signaling	See *register signaling.*
Alternate routing	Deterministic routing arrangement in a switching center whereby traffic overflows to another trunk group when all the circuits in the primary trunk group are busy.
Blocked calls cleared	Calls failing to find idle traffic resources are cleared from the system and do not reappear.
Blocked calls delayed	Calls failing to find idle traffic resources are held in a waiting queue until a traffic resource becomes available.
Blocked calls held	Calls failing to find idle traffic resources are held for an interval equal to one call-holding time. If a traffic resource becomes available during

that interval, it will be used to service the call.

Blocked traffic
Portion of traffic that cannot be handled by a traffic resource (offered traffic minus carried traffic).

Busy hour
Continuous sixty-minute period of the day when the highest usage occurs.

Call
Series of dialing attempts to the same subscriber, where the last attempt results in a completed call or is abandoned.

Call attempt
Any effort, even an error, on the part of a traffic source to obtain service.

Call-attempt factor
Ratio of the number of call attempts to completed calls.

Call-holding time
Length of time during which traffic resources are held for a call.

Carried traffic
Traffic intensity handled by a traffic resource.

Centum-call-seconds
See *unit call*.

Common equipment
Common equipment, or servers, are those resources that are used on a shared basis. Trunk groups, signaling registers, and operator positions are examples of common equipment.

Completed call
Call attempt that is successfully completed (called subscriber answered).

Congestion
See *grade of service*.

Cut through
Circuit is completed between a calling and a called subscriber.

Degree of congestion   Frequency with which a call will encounter some measure of inconvenience in completing a call (i.e., the probability that a call will be blocked or delayed).

Delay system   System in which call attempts are held in a waiting queue until suitable resources are available to handle calls.

DP signaling   Dial-pulse signaling (or rotary-dial signaling) employs a sequence of on-hook, off-hook pulses, at a nominal rate of 10 pulses per second, to represent dialed digits.

DTMF signaling   Dual-tone multifrequency signaling (or tone dialing) uses unique pairs of low- and high-band tones to represent digits.

Erlang   Unit of traffic that represents a circuit occupied for one hour (e.g., one call held for an hour, sixty calls held for one minute each, or any combination totaling one call-hour).

Essentially nonblocking   Grade of service (congestion) that is considered negligible for a specific application.

Final trunk group   Direct trunk group between two switches where alternate routing is not employed.

Finite sources   The ratio of sources to servers is small.

Full availability   Every source has equal access to every server.

Grade of service

Measure of the probability that a percentage of the offered traffic will be blocked or delayed (used synonymously to define level of traffic congestion or service objective).

High-usage trunk group

Trunk group between two switches engineered to carry high traffic loads and offer significant overflow traffic to an alternate route (another high-usage trunk group or a final trunk group).

Hundred-call-seconds

See *unit call*.

Immediate connection

The time required to set up a call is negligible when compared with the associated call-holding time.

Independent sources

Source of calls (subscribers) are independent as far as originating calls are concerned.

Infinite sources

The ratio of sources to servers is very large.

Line signaling

Signaling arrangement, also known as supervision, that controls the setup, holding, charging, and releasing of call connections.

Loss system

System in which a call attempt is rejected when an idle resource is unavailable to handle the call.

MF signaling

Multifrequency signaling, also referred to as MF 2/6 signaling, uses unique pairs of six tones to represent line signals and address digits on interoffice trunks.

MFC signaling

Multifrequency-compelled signaling is similar to MF signaling except that for each pair of tones there is a

	complementary pair that is sent back to the calling switch to verify that the correct signal was received at the called switch.
Network	Arrangement and interconnection of a number of nodes in a telecommunication system, such as the public switched telephone network, corporate private networks, and local area networks.
Nonblocking	Grade of service (congestion) level of zero at all traffic intensities up to and including one Erlang per channel.
Offered traffic	Traffic intensity submitted to a traffic resource.
Overflow traffic	Traffic intensity that cannot be handled by a traffic resource and is offered to (overflows to) an alternate resource.
Peakedness	Abnormal variations in traffic intensity caused by alternate routing overflow traffic or other system variances such as special rates, holidays, floods, and disasters.
Probability	Likelihood of the occurrence of a specific event.
Probability distribution	Probabilities associated with every possible event of a set of events under consideration.
Queuing system	System operating on a delay basis such that when no idle traffic resource is available, a new call arrival waits in queue until a resource becomes available.

Random calls   Calls initiated by subscribers are in-
                                    dependent of calls initiated by other
                                    subscribers.

Register-holding time   See *server-holding time*.

Register signaling   Signaling arrangement, also known
                                    as *address signaling*, that transmits
                                    the called termination's address dig-
                                    its (directory number) through the
                                    network to the terminating switch.

Server   See *common equipment*.

Server-holding time   Length of time during which a traffic
                                    resource is held to process a call
                                    attempt.

Service objective   See *grade of service*.

Traffic density   Number of simultaneous calls at a
                                    given moment.

Traffic intensity   Average traffic density (occupancy)
                                    during a one-hour period.

Traffic flow   Product of the number of calls dur-
                                    ing a period of time and their average
                                    duration (call-holding time).

Traffic resource   Any resource provided to handle traf-
                                    fic, such as trunks, signaling regis-
                                    ters, and switching matrices.

Unit call   Unit of traffic, also known as centum-
                                    call-seconds (CCS) or hundred-call-
                                    seconds, representing a circuit occu-
                                    pied for one hundred seconds.

Glossary

BASIC	Beginner's All-purpose Symbolic Instruction Code (programming language)
CCITT	International Telephone and Telegraph Consultative Committee
CCS	centum-call-seconds (or hundred-call-seconds
CO	central office
DP	dial pulse
DTMF	dual-tone multifrequency
Erl	Erlang
FIFO	first-in, first-out
FORTRAN	FORmula TRANslation (programming language)
GOS	grade of service
MF	multifrequency

MFC	multifrequency compelled
PBX	private branch exchange
PSTN	public switched telephone network
UC	unit call

Selected Bibliography

D. Bear, *Principles of Telecommunication—Traffic Engineering*. London: Institution of Electrical Engineers, 1980.

P. Beckmann, *Elementary Queuing Theory and Telephone Traffic*. Geneva, IL: abc TeleTraining, 1977.

V. E. Benes, "On Trunks with Negative Exponential Holding Times Serving a Renewal Process," *Bell System Technical Journal*, vol. 38, no. 1, 1959.

G.S. Berkeley, "Traffic and Delay Formulae," *J. Post Office Elec. Engineers* (London), vol. 29, 1936.

J. G. van Bosse, "Delay Distribution in Communication Systems with Partially Ordered Queues," *IEEE Trans. Commun. Sys.*, vol. 9, no. 3, 1965.

J. R. Boucher, *Voice Teletraffic Systems Engineering*. Norwood, MA: Artech House, 1988.

E. Brockmeyer, H. Halstrom, and A. Jensen, "The Life and Works of A.K. Erlang," *Trans. Danish Acad. Sci.*, no. 2, 1948.

G.A. Campbell, "Probability Curves Showing Poisson's Exponential Summations," *Bell System Tech. J.*, vol. 2, no. 1, 1925.

J. W. Cohen, "The Generalized Engset Formula," *Philips Telecommun. Rev.* (Amsterdam), vol. 18, no. 4, 1957.

C. D. Crommelin, "Delay Probability Formula When the Holding Times Are Constant," *J. Post Office Elec. Engineers* (London), vol. 25, 1932.

C. D. Crommelin, "Delay Probability Formulae," *J. Post Office Elec. Engineers* (London), vol. 26, 1933.

A. Elldin and G. Lind, *Elementary Traffic Theory*, ed. 2., Stockholm: L. M. Ericcson, 1964.

A. K. Erlang, "Solution of Some Problems in the Theory of Probability of Significance in Automatic Telephone Exchanges," *J. Post Office Elec. Engineers* (London), vol. 10, issue 4, 1917.

T. C. Fry, *Probability and Its Engineering Uses*, ed. 2., New York: D. Van Nostrand, 1965.

R. Holm, "The Validity of Erlang's Trunk Congestion Formula," *J. Post Office Elec. Engineers* (London), vol. 17, 1925.

M. M. Jung, "Loss Probability Charts Calculated with the Formula of Engset," *Philips Telecommun. Rev.* (Amsterdam), vol. 23, no. 4, 1962.

G. Lind, "Accuracy of Traffic Measurements in a Poisson Traffic Process," *Proc. IEE* (London), vol. 23, no. 4, 1967.

E. C. Molina, "Application of the Theory of Probability to Telephone Trunking Problems," *Bell System Tech. J.*, vol. 6, no. 3, 1927.

E. C. Molina, *Poisson's Exponential Binomial Limit*, New York: D. Van Nostrand, 1942.

I. Molnar, *Delay Probability Charts for Telephone Traffic Where the Holding Times Are Constant*, Chicago: Automatic Electric, 1952.

J. Riordan, "Delay Curves for Calls Served at Random," *Bell System Tech. J.*, vol. 40, no. 3, 1953.

J. Riordan, "Delay for the Last Come First Served Service and the Busy Period," *Bell System Tech. J.*, vol. 38, no. 4, 1951.

R. Syski, "Determination of Waiting Time in the Simple Delay System," *Automatic Electric Tech. J.*, vol. 13, 1957.

R. Syski, "The Theory of Congestion in Lost-Call Systems," *Automatic Electric Tech. J.*, vol. 9, 1953.

Telephone Traffic Theory Tables and Charts, Munich: Siemens AG, 1970.

Teletraffic Engineering Manual. Stuttgart: ITT Standard Electrik Lorenz AG, 1966.

R. I. Wilkinson, "Working Curves for Delayed Exponential Calls Served in Random Order," *Bell System Tech. J.*, vol. 32, no. 2, 1953.

Index